MW01002554

A Year in the Notch

A Year in the Notch

Exploring the Natural History of the White Mountains

William Sargent

UNIVERSITY PRESS OF NEW ENGLAND | Hanover and London

University Press of New England, Hanover, NH 03755

Printed in the United States of America

5 4 3 2 1

Library of Congress Cataloging-in-Publication Data

Sargent, William, 1946–
A year in the notch : exploring the natural history of the White Mountains / William Sargent.
p. cm.
Includes index.
ISBN 1–58465–011–7 (cloth : alk. paper)
1. Natural history—White Mountains (N.H. and Me.) I. Title.
QH104.5W48 S27 2001
508.742'2—dc21 00–011954

All photographs are by William Sargent unless otherwise noted.

Acknowledgments for poetry quoted:

Excerpts from "In Hardwood Groves" and "The Need of Being Versed in Country Things" by Robert Frost from *THE POETRY OF ROBERT FROST* edited by Edward Connery Lathem, Copyright 1923, 1934, © 1969 by Henry Holt and Co., copyright 1951, 1962 by Robert Frost. Reprinted by permission of Henry Holt and Company, LLC.

Excerpt from "On the Porch at the Frost Place, Franconia, NH" from *SELECTED POEMS AND TRANSLATIONS 1969–1991* by William Matthews. Copyright © 1992 by William Matthews. Reprinted by permission of Houghton Mifflin Company. All rights reserved.

Frontispiece: Franconia Notch.
Dedication: Chappell emerging.

To my daughter Chappell, a far better writer than I!

Contents

Part IV *Summer*

Part V *Autumn*

Preface

We shall not cease from exploration,
And the end of all our exploring,
Will be to arrive where we started,
And know the place for the first time.
—T. S. Eliot, *Four Quartets*, 1943

This book, written about a place so close to home, came about because of trips to places far away. The first was to investigate volcanoes on Iceland and Montserrat. The islands seethed with the heat of first creation. They belched life-forming carbon. They recreated conditions at the very beginnings of our planet; yet they represented but an instant in time, a snapshot of our earth's beginnings. I had traveled to the islands to write a book about volcanism, but gradually I realized that a far more profound story lurked within the rocks of my own backyard.

The rocks of northern New England have lived many lives. They were born 500 million years ago on the sandy floor of a tropical sea. They drifted on a crustal plate from the equator to close to the Arctic Circle. They were

Soufriere Hills volcano, Montserrat.

blown out of volcanoes, thrust up as mountains, worn down by erosion. Only in the most recent Ice Age have they been polished by glaciers and overgrown by trees.

Our rocks have seen the Atlantic Ocean open, close, and open again; they have witnessed almost three-quarters of our earth's history. Their story may lack the immediate drama of volcanoes and dinosaurs, but it reveals the subtle evolution of our nuanced planet. I had intended to write a page-turning potboiler but the story of these stones called for the talents of a Russian novelist. Was I up to the task?

One more trip sealed my fate. This trip was to the Rocamador pilgrimage near the Dordogne Valley of south central France.

It was a wet, cold, foggy day when we finally arrived at the sacred spot. Tendrils of mist curled off the Alzou River and climbed up steep limestone cliffs to entwine themselves about the battlements of an ancient gray fort that looms above the valley.

A grand stairway twists through a town of tourist shops before climbing to the cathedral overhead. Medieval penitents had to climb the 1,400 stone stairs on their knees in order to gain redemption. At every switchback, they would pause at a small shrine that reminded them of the twelve stations that marked Jesus' journey to his crucifixion. By the time they reached the summit, they were walking on the exposed bones of their naked bloody knees.

As I trudged up the cliff, I stared at the steps trying to imagine the scene. But I was suddenly distracted by the sight of fossils. Yes, embedded within the limestone stairs were fantastic, elegantly beautiful fossils. Some were of shells, others were of bits and pieces of fish, but the most exquisite were of some sharp, fluted, arrowhead forms.

My imagination raced. Millions of years ago, these cliffs were on the bottom of a shallow Jurassic sea. It teemed with billions of marine animals. Could these fossils be the pens of squid that performed an elaborate courtship before mating and dying by the millions?

I pulled shopkeepers out of their shops and guides off the street. But nobody seemed to know what the curious arrow-shaped forms were. One

Squid fossil, Rocamador Pilgrimage.

The Old Man of the Mountain.

shopkeeper displayed both the Gallic pride and the Gallic disdain that one expects to pay for so dearly in France, "Mais oui, but your so-called fossils are nothing more than the slime trails of escargot."

The guides were no more helpful. They could tell me every last detail of the spate of miracles that were supposed to have occurred at this site; yet they could tell me nothing of these fragile animals. For me, one set of miracles reflected on the folly of our own species; the other, the wonder of life preserved forever in limestone.

The pilgrimage was created after the discovery of the grave of an old mystical hermit. He lived in a cave in these cliffs where he "loved the rocks" and worshiped nature. Yet over time, the people who developed this site turned away from the worship of the earth to the worship of their own blood-stained past.

Today, we too worship ourselves, our present, our technology. We have almost lost the ability to worship the earth, rocks, trees, and mountains; the only things that make it all possible. In writing this book I hope I have rekindled some of that ability to draw wisdom, awe, and inspiration from the earth . . .

Franconia, N.H.
June 2000

W.S.

Acknowledgments

Many people have helped me write this book. The first were my parents, who had the foresight to buy a farmhouse near Franconia Notch. The small red house sits on a westerly slope overlooking a long narrow meadow. A mountain brook bubbles out of a shaded spring above the house. It gurgles out of the forest, swirls around our barn, then tumbles through the gloriously weedy meadow to the Gale River, the Connecticut, and the Atlantic Ocean beyond.

In the spring, fiddleheads push through the moist soil to unfurl as towering copses of cinnamon fern. On June nights, the meadow hosts scores of fireflies that drift lazily from tree to tree. In late summer, ruby-throated hummingbirds dart, dip, and hover over their favorite patches of jewelweed. Bell-like "tinks" reveal that the feisty birds are deadly serious about defending their feeding territories.

The brook is often the focus of our summertime projects. Once we dammed its upper reaches to make a glade complete with water flowers and birch-log benches. Another time, we collected slate to make a series of splash rocks that could be heard from our bedroom window. Last summer, we pried out boulders to make pools to hold our palustrine treasures.

Over the years, I have made a habit of enjoying my morning coffee sitting on the porch that overlooks the tree-lined meadow. The former field leads down the hill, across the brook, and back up another rise to an ancient apple tree. I have watched deer, turkey, coyote, and bear looking back at me from the far end of the meadow. They seem comfortable there, knowing they can leap into the safety of the deep forest with just a few long bounds.

But most importantly, the house became a base for my rambles through northern New England. Many neighbors have helped me with these excursions. We first met the elegant Joanne Balch teetering on the top of a high ladder. She came down to tell us about the bears of Coal Hill. Later we met Chris Balch, who introduced me to the fine art of discovering moose antlers. He remains at least five antlers ahead of me. Jim and Evie Fitzpatrick live on the Gale River. They have introduced us to turkeys, told me their stories, and provided much-needed transportation.

Mrs. Otterowski owns the beautiful house at the top of Coal Hill. It boasts a pond, a moose yard, and gorgeous natural gardens. It is as perfect

Opposite: *Ferns in the meadow.*

Sunlight in the stream.

As perfect as a house could ever be.

Cannon Mountain.

as a house could ever be. She and Kurt Hohmeister have been unceasingly generous with their love and knowledge of the land. Ritch MacLaughlin and Ed Clough have shared their many wonderful memories of the woods. Anne de Rham and Letty Crosby have provided food, good cheer, animal companionship, and entertainment.

It is said that if you learn to ski on Cannon Mountain you can ski anywhere on earth. It remains a skier's mountain; happily unprettified by shops, lodges, and condos. The personnel of the mountain have kindly provided transportation in the best and worst of weather. Likewise, the people who run the Mount Washington Auto Road and the Mount Washington Cog Railway have been extremely generous. I think often of Sarah Curtis and her dedicated crew who woman the Mount Washington Weather Observatory. It seems unfair that I spent three of the warmest days on record at the summit, while they endure never-ending winters on the rockpile. I will wager, however, that they are grumbling about the warmth and lack of snow as this new millennium unfolds.

Chris Martin kindly allowed me to join New Hampshire Audubon's annual trip to band peregrine falcons in Franconia Notch. Phil Lovejoy of the Harvard Museum of Natural History invited me to join their field trip to the Palermo Mine. Kathy Fallon Lambert, of the Hubbard Brook Ecological Foundation, and all the people of the National Forest Service could not have been more knowledgeable and accommodating during my investigations of Hubbard Brook.

I spent a wonderful day with Todd Bogardus of the New Hampshire Fish and Game Department. I'm sorry I couldn't tell him then that I had already decided to tell the story of bearbaiting from the bear's point of view! The Robert Frost Place put up with my prosaic style while providing their poetic inspiration.

In the interests of full disclosure, I would also like to admit that I have

Our neighbor's mailbox.

taken some literary license in trying to recapture the mood and language of some of the people in this book. The chapters that take place at the Hubbard Brook field station and the Village Inn Bar provide examples. The conversations I relate never took place all in the same room at the same time. They are distillations of similar conversations I had on several visits to these fine establishments.

As a science writer, I have discovered that often the most effective way to find out what a scientist really thinks is to get him out of the lab for an "over a beer," off-the-record conversation. It is, of course, a strategy that political writers have used for years.

So, if anyone takes issue with the veracity of my conversations, my only defense is that I'm sure they are remembered at least as accurately as similar conversations at either of these two venerable institutions. Besides, the lights were out, I wasn't taking notes, and a lot of beer was being consumed.

Penultimately, I would like to thank Kristina and Chappell, who shared many adventures, but also had to endure my writing about them. Finally, I would like to thank all the people mentioned in the text. You were unceasingly patient and are hereby released of responsibility for any flatlander mistakes. They are mine and mine alone.

PART I

Beginnings

The only fault I find with New Hampshire, is that her mountains aren't quite high enough.

—Robert Frost, "New Hampshire"

Mount Washington.

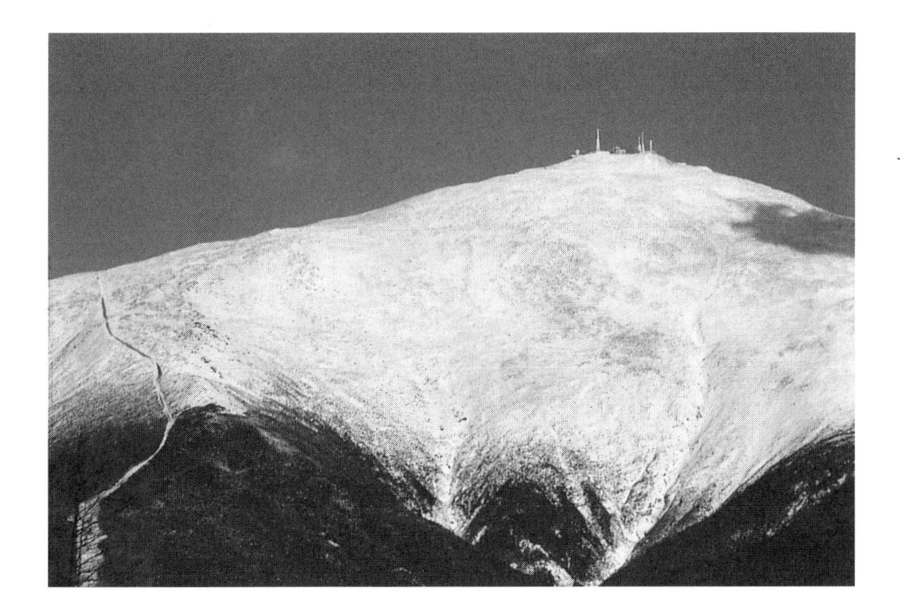

The granite state's most revered summit, Mount Washington, is fashioned from schist, not granite.

Chapter 1

The White Mountains

Two hundred miles north of Boston is a world of craggy mountains, boreal wildlife, and the worst weather on earth. For generations, this world has provided escape, transcendence, and a touch of the sublime.

These are the White Mountains of northern New England. They include Franconia Notch to the west, the high mountains of Vermont and New Hampshire to the north, the Presidentials and the Mahoosuc Range of Maine to the east. Most are built upon massive blobs of frozen granitic magma. But the Granite State's most revered summit, Mount Washington, is fashioned from schist, not granite!

To the west of the White Mountains are the Green Mountains of Vermont, the Taconics of New York, and remnants of the Canadian Shield. To the east are the New England uplands and the blue-green glimmer of the Atlantic Ocean.

From their summits, these mountains have a similar feel. They stretch north to south, they thrust above the surrounding land, they are snow-capped and mantled with green. The highest possess an otherworldliness, with seas of ice-cracked boulders, winds of unrelenting force, meadows of fleeting alpine beauty.

Yet deep within their cores, these mountains tell another story. Some are over a billion years old, others were once under the sea, some sparkle with garnets and sillimanite, others are masses of stolid granite.

For years, geologists had no way to explain these mountains, but that did not stop them from spinning yarns. Uplifts and thrust faults explained the ranges, transgressions and regressions of the oceans explained why marine fossils were found on mountain tops.

But that all changed in the 1960s. Scientists rearranged the clues to arrive at a new understanding. The surface of the world was not a static entity; it could move in ways unimaginable only decades before. Plate tectonics was truly a revolutionary new concept. In earlier times, architects of such a radical idea would have been excommunicated if not burned at the stake. Today they are given Nobel prizes.

I had the good fortune to witness this revolution from the safety of Harvard College. The university has a venerable tradition of being on the wrong side of almost every revolution in the past three hundred years. My particular perch was Harvard's introductory geology course, nicknamed "Rocks for Jocks." True to form, the department was split down the middle between true believers of the old paradigm and heretics of the moving earth. I happily fell in with the latter. It was like being in London when Darwin was proposing evolution or Vienna when Freud was laying the foundations for psychoanalysis.

Scientists suddenly had a new way to understand why there are earthquakes, volcanoes, mountains, and oceans. These things were relatively easy to understand in the Himalayas or on the Mid-Atlantic Ridge, but they are damn hard to interpret in our own backyard. Now we know that northern New England has had a billion years of oceans opening and closing, islands crushing into mainlands, and continents leaving little bits and pieces of themselves glommed on to our shores.

So, before we turn our attention to the story of northern New England, it will help to take a brief detour into the excitement of plate tectonics. Briefly stated, the theory of plate tectonics says that the earth's crust is made up of twelve major plates and about twenty minor plates that carry continents and oceans on their backs. The plates are pushed apart at mid-ocean ridges and plunge back into the earth's interior when they collide with other plates.

The Mid-Atlantic Ridge provides an example. It is in the process of pushing apart the American and European plates at the rate of about six inches per year. You can see this process without getting your feet wet by traveling to Iceland. Iceland is the only major island that sits on the Mid-Atlantic Ridge above the surface of the ocean. The center of the island is bisected by a huge rift valley.

Every so often, mile-long fissure volcanoes light up the evening sky. Magma from the eruptions hardens to become part of the European plate that makes up the eastern side of the island or the American plate that makes up the western side of the island. As more magma comes to the surface in the middle of the island, it pushes the plates aside, so that the island is growing at the rate of several centimeters per year.

If a plate is being pushed very fast, it will plunge under another plate and

Fissures open on Iceland as the American plate moves west and the European plate moves east.

melt as it reenters the earth's interior. The melted plate rises as huge tear-shaped blobs of molten magma. When the magma reaches the surface, it erupts as an andesite volcano. The magma that remains below the surface may harden as granite or pegmatite. This process can presently be seen in the Caribbean, where volcanoes are building islands such as Montserrat and Martinique. Eventually, plate tectonics will push this arc of islands to the west and crumple them into the mainland. They will become part of the continent being formed at the Isthmus of Panama.

When the plates come together, they push ocean sediments and island arcs against continents, then fold them all into mountains. The Himalayas are presently undergoing such a mountain-building orogeny, which is thrusting up Mount Everest at the rate of an inch per year.

After plates have come together, the process starts all over again. Heat rising from the earth's interior rifts apart the continents, water flows in, and a new ocean is born. This process is presently happening in the great rift valley of Africa and below the Red Sea. Part of the eastern section of Africa will soon drift out into the Indian Ocean in the next million years or so.

As the ocean forms, it leaves behind a continent of eroding, weathering mountains; mountains embedded with the remains of the volcanics, sediments, and exotic terrain of its violent formation.

Over the past billion years, New England has undergone all these changes. Originally, our plate was a protocontinent several hundred miles south of the equator. The remains of that original continent is our present

Canadian Shield. As ancient plates converged, New England experienced the Taconic, Acadian, and Alleghenian orogenies; multiple orogenies if you will, that created first the Taconic Mountains of New York, then the Green Mountains of Vermont, then the arc of once-tropical islands that nestles beside the present Connecticut River valley, then the White Mountains of New Hampshire, and finally the exotic terrain of the New Hampshire coast, once a part of Africa. In that one long sentence, New England moved from the equator to its present position, and created and eroded three mountain ranges once the height of today's Himalayas.

How did the earth do it in only a billion years? That is a story for "Gathering Slate."

Wisps of fog coil through the conifers of Franconia Notch.

Chapter 2

Gathering Slate

NOVEMBER 14, 1998

It is a long, cold, gray day in early November. The peaks of Cannon and Lafayette are still enshrouded beneath lenticular clouds. Wisps of fog coil and uncoil through the silent conifers of Franconia Notch. Robert Frost's children thought the fog looked like Chinese dragons evolving and devolving from mist and water. Frost stole the metaphor but gave credit where credit was due.

The landscape looks almost as bleak as when the glaciers retreated from this area ten thousand years ago. The first snow has yet to fall, the leaves are gone, rocks are still prominent. It will be a good day to explore back in time, a billion years before the glaciers put the finishing touches on this ever-changing landscape.

To do so, I will drive to the Vermont border twenty miles away. The reason I give my family is that I am going to gather some slate to line our garden path. Of course, that is but an excuse. My real reason is that I want to travel back in time to when this part of our continent lay underwater.

Each mile west carries me back millions of years. The glacial erratics were deposited in the Notch ten thousand years ago. The pink granites of Cannon Mountain squirted through the existing bedrock 150 million years ago. The Littleton schists that underlie Mount Washington are 350 million years old, but they were metamorphosed from shale that was already 450 million years old.

Crossing the short span of the Connecticut River plunges me back 50 million years more. The rocks on the New Hampshire side are from 450-million-year-old volcanic islands, the slate on the Vermont side is from 500-million-year-old shallow seas. The core of the Adirondack Mountains are over one billion years old.

In the early summer you can see the difference. The open fields of Vermont seem somehow greener than those of New Hampshire. This is not just an illusion sponsored by the Montpelier Chamber of Commerce. The grass really is greener in Vermont, nourished neither by the fervent wishes of its citizenry nor by the overlying metaphysics of the singular state, but rather by its underlying bedrock. Grass grows greener on limestone-rich sedimentary rocks than on acidic igneous granite. That is the reason Vermont's main crop is milk—New Hampshire's, trees.

Such musings make the trip pass quickly. Soon I am at my destination. An outcrop of slate rises above the Vermont side of Interstate 93. The dark slate shimmers with flecks of mica and is studded with garnets the color of wine. The foliation is almost vertical, but it is easy to see that the beds once lay on the bottom of a shallow tropical sea that stretched over what is now New Hampshire, Vermont, and Maine. These rocks are clues to a complex but fascinating story.

Four hundred and fifty million years ago, this part of the earth's crust was under a shallow sea hundreds of miles south of the equator. Trilobites scuttled over dark volcanic muds and burrowed in the sediments to shed their shells. Ten-foot-long Eurypterids slithered over the bottom capturing the jawless fish of this mid-Ordovician sea. On the continents, dragonflies with four-foot-long wingspans skimmed over a landscape of mosses, ferns, and horsetail plants.

The sea-floor muds came from sediments flowing off an arc of active vol-

The Connecticut River. The grass really is greener in Vermont, but it's because of limestone, not the fervent wishes of its citizenry.

Crinoids in Ordovician slate.

canic islands hundreds of miles to the east. Today, the cores of those islands make up the Bronson Hill Formation, a group of unassuming hills, just across the Connecticut River in New Hampshire. How did those islands move six hundred miles to the west?

During the 1970s, scientists tackled that question, and the Bronson Hill Formation became a crucial piece in fitting together the global jigsaw puzzle called plate tectonics.

Geologists knew that 450 million years ago two major plates were converging, closing what we can call the early Atlantic Ocean. In fact, it was actually a former ocean that preceded the Atlantic by hundreds of millions of years. The force of this ocean's closing broke off a small piece of sea floor plate. The major plate continued to plunge under the new plate, melting and rising as magma to create an arc of volcanic islands six hundred miles from the mainland.

If you want to see a similar scene, all you have to do is fly to the modern Caribbean. There, the arc of islands that curves gracefully from the Virgins to the Grenadines overlies the most deadly area of volcanism on the earth. The islands lie on the Caribbean plate, a small plate that broke off the Atlantic plate about a million years ago. Today, the Atlantic plate plunges beneath the broken Caribbean plate at an alarming rate. As it descends back into the earth's interior, the Atlantic plate melts and huge inverted teardrops of molten magma rise toward the surface. They emerge as deadly volcanoes.

The most recent Caribbean volcano to erupt was on Montserrat. The most deadly was on Martinique, where thirty thousand people lost their lives in forty-eight seconds. In a hundred million years, the plate carrying these islands will be pushed a thousand miles west to be folded into Central America building towering mountains into that growing proto-continent.

How did these mountains move 600 miles?

The same thing happened in modern-day New England. During the Taconic mountain building cycle, plate tectonics pushed the volcanic islands up against Vermont, folding 600 miles worth of islands and sediments into mountains the height of today's Himalayas. The thick beds of lava that used to lie on top of the islands eroded away, leaving only the plutonic cores of the islands behind. Today they make up the Bronson Hill Formation, a narrow arc of hills that snuggles up beside the Connecticut River along the length of the Vermont–New Hampshire border.

If geologists were naming the Bronson Hill Formation today, they would be hard pressed not to call it the Garnet Hill Formation, after the neighboring hill that gave its name to the catalogue company. Do the clothing manufacturers know that their brand-name hill was once a tropical island, not a snow-covered New Hampshire mountain?

After the Taconic event, plate tectonics calmed down, as the ocean continued to close uninterrupted by any more pieces of broken crockery. Finally, 350 million years ago, the early Atlantic completely closed, crushing Africa up against America to form the supercontinent of Pangia. This was the Acadian orogeny that lifted a second range of mountains the height of the present-day Himalayas. These mountains also eroded back to their metamorphosed cores.

One of these cores is my next stop, Mount Washington. Here, I find some of the same slate I found in Vermont, but it is almost unrecognizable. During the Acadian orogeny, the slate was buried under seven miles of earth and cooked by the heat of the collision of the two continents. The result was Littleton Schist, a hardy rock laced with seven-inch-long needles of greenish-yellow sillimanite. Today many people use blocks of this beautiful rock to make stunning front door steps.

My final stop is back at Cannon, one of the youngest mountains in the area. It was formed two hundred million years ago when Pangia started to rift apart again, opening the present-day Atlantic Ocean. As it did so, huge faults developed and magma squirted through former formations.

The magma crystallized as the pink Conway granite now used in monumental buildings throughout America. It also formed the bedrock of the Flume, the Basin, and Cannon Mountain, with its craggy profile of "The Old Man of the Mountain." The pluton was probably helped by a hotspot, a deeper plume of magma that remained stationary as the plate slid over it. Today the same hotspot is believed to be pushing up the Madeira Islands!

As the modern Atlantic Ocean opened, it left behind a significant portion of the former African continent still attached to the American continent. Today, we call this exotic terrain eastern New Hampshire. It also created what geologists call an unconformity, an area where rocks from different ages and places lie against one another. Such a place is on the Connecticut River. The rocks on the Vermont side are part of the ancient craton and shallow seas that made the American continent. The rocks on the New Hampshire side came from oceanic islands and Africa.

It is with considerable satisfaction that I drive back home. Within twenty miles I have seen the results of one ocean collapsing, two continents colliding, and another ocean being born, over a billion years of history. It has been a long trip through one of the most fascinating but complex areas in the world.

Just above Echo Lake, I find a large piece of granite that has fallen off an outcrop. A beautiful vein of younger white pegmatite has melted its way through the darker granite of Mount Lafayette.

When I get home, I haul the piece of heavy, white New Hampshire granite into the garden and place it on a pedestal of dark gray Vermont slate. I have created my own 500-million-year unconformity. And, of course, I can finally tell my family why the grass is always greener in Vermont. It has been a worthwhile day.

Cannon Mountain. Her granite is exfoliating.

Tuckerman's Ravine. "Glacial epochs are great things, but they are vague, vague."

Chapter 3

"On the Rocks"

Glacial epochs are great things, but they are vague, vague. —Mark Twain

NOVEMBER 27, 1998

"Why did I ever agree to write this book," I mutter to myself as I drive north on Route 93. Naked, lifeless trees stand on either side of the darkened highway. It is late November, halfway between the exuberance of autumn and the white clarity of winter; the gray season in New Hampshire.

"I don't like winter. I have seasonal affective disorder. Last year I had the good sense to fly to Montserrat to write about volcanoes, which I will take over snow any time of day . . ."

My kvetching is suddenly interrupted. Motes of tell-tale white hurtle out of the blackened sky, then swirl around my solitary car. I look again. "Snow, good God, snow!" Billions of flakes of snow dance and flicker in the beams of my headlights. My mood soars as the unexpected gift cavorts around my cozy automobile. Soon the snow is accumulating on the side of the road, weighing down the boughs of evergreens.

Blue lights flash ahead. Five police cars shoot past in the opposite direction. A car behind us has overshot the shoulder and plunged down the embankment. Above us, a pair of hikers are digging an emergency snow enclosure. The sudden storm caught them off guard. Now they will have to spend the night several feet from the top of Mount Lafayette.

Ahead of us is a grim reminder of just how ruthless this mountain weather can be. The exit sign says Boise Rock. It was named for an early traveler who also became trapped on Mount Lafayette. Boise sought shelter beside an erratic boulder, shot his horse, and crawled into its steaming entrails for the night. It has always seemed more appropriate to have called the rock, "Boise's Horse's Rock," but be that as it may. Evidently George Lucas once heard the story, for he used it to good effect in his Star Wars movies.

But tomorrow the storm will be over. Yuppies will be complaining about the power surges that fried their microchips and VCRs. We forget we still live in the grip of a three-million-year-old Ice Age.

Only an eyeblink back in geological time, this valley was encased in ice two miles thick. Even the tops of Lafayette and Cannon Mountain didn't pierce through the featureless expanse of ice and snow.

The process started when the earth's orbit, tilt, and wobble conspired to cool the planet. But it was pushed over the edge by two sets of mountains. The first were the newly emerged Himalayas. Their height had cooled the earth by altering the world's jet streams. More importantly, their weathering had cooled the earth by removing carbon dioxide from the atmosphere and sequestering it as plankton in the ocean. The second mountains were the volcanoes of central America.

Twenty million years ago, the Pacific plate, a huge piece of the earth's crust, plunged below the Caribbean plate, crumpling the seafloor above the surface and injecting it with volcanic lava. By three million years ago, a string of volcanoes linked North and South America; the Isthmus of Panama had been born.

The Isthmus cut off the warm equatorial currents that formerly encircled the world. Now the currents were deflected north, transporting tons of warm moisture that evaporated into the atmosphere of North America. Storm clouds dropped this moisture as rain and snow on the mountains of North America.

Above us on Mount Lafayette, the process is being repeated. Moisture-laden air blowing in from the west is being deflected up the side of the mountain, where it cools. As it cools, it drops its load of water as heavy snow. In some winters, the ravines of high mountains such as Lafayette accumulate as much as fifty feet of snow and the snowpack remains until well into June.

All that it would take for another phase of glaciation to start would be for the snowpack to remain in the mountains for three more months, and then grow again in winter. This started to happen in 1816, when the Tambora volcano lofted tons of cooling ash into the atmosphere. New England had a year without a summer, known locally as "Eighteen hundred and froze to death."

When snow caps become more than a mile thick, the ice beneath them becomes plastic and starts to flow. That started happening about three mil-

Mount Lafayette. "Sometimes the snowpack remains until well into June."

lion years ago, initiating the Ice Age, a series of glacial and interglacial periods that continues today. Technically, we are still in an interglacial that is supposed to last for another two thousand years or so.

The last glacier to push through Franconia Notch was the Laurentide glacier. It expanded from northern Canada, advancing through the Notch less than eighty thousand years ago. A wall of ice a quarter- or half-mile high surged through this valley at the rate of two football-field lengths per year. Tons of ice tore away the sides of mountains, scored the bedrock, and scooped out magnificent bowls or cirques from the sides of mountains. The glacier slid past either side of a Mount Lafayette, sharpening its summit into the knife-edged ridge that our climbers had so wisely abandoned.

At the peak of its advance, about eighteen thousand years ago, the glacier even scraped over the tops of these mountains. If you had been able to fly over it, all you would have seen was a featureless expanse of white extending as far south as the Midwest and New York.

Then something changed. The Millankovich cycles that determine the earth's tilt, axis, and wobble came back into synch. The glacier started to melt more in summer than it grew in winter. In its wake, the glacier left a jumble of gravel, boulders, and sand similar to what you might see below glaciers in Iceland or Alaska today.

Where the leading edge of the glacier melted, it dropped great parabolic arcs of till. Today, these terminal moraines buttress the islands of Martha's Vineyard and Nantucket. Streams and rivers flowing under the glaciers deposited sinuous rows of sand and gravel. These eskers crisscross New England, providing a ready supply of clean building material. Some of the most impressive eskers are those along Interstate 93 just south of Franconia Notch. Below the glaciers, conglomerates of mud, sand, and clay got wadded up and left behind as drumlins. Today, one of the best-known drumlins is Charlestown's Bunker Hill, which proved to be such expensive real estate to British troops during the American Revolution.

Just to the south of Hartford is Rocky Hill, Connecticut, site of a terminal moraine that dammed up Lake Hitchcock, an ancient glacial lake that stretched two hundred miles north into Canada. A finger of the lake even jutted rather rudely into the town of Franconia. Today, the Connecticut River Valley lies in the bed of the former lake and its former sediments provide the only easily tillable soil in New England.

The glacier melted back through Franconia Notch at the rate of about four football fields per year. In its wake, it left a beautiful U-shaped valley. Artist's Bluff helped slow the glacier's retreat by dumping a load of sand and gravel that today dams up Echo Lake. Waters trickling out of the east side of the lake become the Pemmigewasset River, which joins the Merrimack to flow into the Atlantic Ocean north of Boston. Waters flowing out of the west side of the lake flow into the Gale River and join the Connecticut River to flow into the Atlantic Ocean in Long Island Sound.

Immediately after its retreat, the glacier left a desolate scene in Franconia Notch. The topsoil had been swept away; only a desolate jumble of rocks, boulders, sand, and gravel remained. Yet that changed with remarkable speed. Ten years later, tundra vegetation—mosses, sedges, and grasses—had spread down from the mountain tops to dot the land. Fifteen years later, rain had leached the alkalines out of the surface, and tundra vegetation had added organic matter to the soil. Seedlings of conifers started to sprout amidst the thickets of tundra vegetation.

Seventy-five years later, the spruce and fir trees had grown so high that they shaded their offspring's growth and the shade-tolerant hemlock were already starting to sprout in the understory. Franconia was starting to look much like it does today.

Elephant Head. This feature was polished by a glacier as it passed through Crawford Notch.

Though no remains have been found, cave bears, mastodon, and musk ox undoubtedly passed through Franconia Notch. Perhaps beavers the size of bears were damming up ancient rivers or beginning to evolve that complex new behavior. The reason we have not found the remains of paleohumans in the White Mountains is that they probably avoided the mountains as their Indian descendants did during historic times. We do know that their counterparts were hunting many of the species of large paleolithic megafauna to extinction, leaving open niches that have yet to be refilled.

We can also imagine what happened to these cold-adapted animals during the Younger Drydas, a fascinating little blip in the climatological record. So much fresh water was flowing into the Atlantic Ocean from the melting glaciers fifteen hundred years ago that it made it impossible for the surface waters to plunge toward the bottom. Normally, cold, dense, saline-rich water plunges downward, initiating bottom currents that transport the world's cold arctic waters to the equator for warming.

These thermohaline currents stopped, then started again, first cooling, then heating the world. In as little as fifty years, the average annual temperature rose seven degrees in the north and four degrees worldwide. Summer must have seemed unbearably hot to the cold-adapted tundra species. Per-

Glaciers left this U-shaped valley in Crawford Notch. Courtesy *The Courier* newspaper.

Modern moose keeping cool. W. J. Otorowski and Betty Otorowski.

haps they migrated to the tops of mountains the way moose do today. Perhaps the rapid changes helped lead to their demise.

It is something to ponder as we face a world that seems to be warming at the rate of one-tenth of a degree per decade. That modest rate is already raising sea levels, increasing hurricanes, disrupting agriculture, destroying homes, and possibly causing El Niños and La Niñas. What will happen if human-induced warming alters the world's ocean currents the way they were disrupted by the Isthmus of Panama or during the Younger Drydas period? Would it trigger another Ice Age, a runaway greenhouse effect, or a world fluctuating madly back and forth between the two extremes? Perhaps tomorrow's weather will give us some clues.

PART II

Winter

There may be worse weather, from time to time, at some forbidding place on the Planet Earth, but it has yet to be reliably recorded.

—William Lowell Putnam,
Defending Mount Washington's Weather, 1991

The building where Sal Pagliuca recorded "the worst weather on earth."

Franconia Notch.

Chapter 4

First Snow

NOVEMBER 28, 1998

A strange, deep silence envelops Franconia Notch. It is the profound quiet of newly fallen snow. It muffles every sound and makes the silhouettes of trees stand out in stark relief along the ridges of distant mountains. Wisps of snow blow off the peaks of Cannon and Lafayette shimmering in the light of the early morning sun.

All that I can hear is the sound of my own breath as I labor up the side of Artist's Bluff. It is surprisingly difficult to find footing on the icy rocks beneath the snow. Perhaps I am the only person on the mountain. I hope that yesterday's climbers have awoken on Lafayette and are now digging out of the sixteen inches of newly fallen snow.

I scan the horizon. Nothing. The mountain air startles me with its clarity. I have become used to the softer lines of coastal air and the constant haze of city pollution. Sometimes you can spot the breath of a bear hovering over its hibernation den. But today's temperature is already too warm. Even the surface of Echo Lake is only partially frozen.

A single, perfect snowflake lands on the sleeve of my parka, then melts quickly away. It transports me back to a night in college when I was aimlessly perusing the stacks instead of studying for exams. My glance fell on a book called, simply, *Snow Crystals*, by Wilson Bentley.

Little did I know that *Snow Crystals* had once been New England's answer to Monet's *Water Lilies*. Bentley was a Vermont farmer had who devoted his entire life to photographing snow. He would go out every night to catch snowflakes on frozen slides, then quickly photograph them under his microscope before they melted. His collection of starkly beautiful black-and-white photos became a nineteenth-century sensation, earning him the sobriquet "Snowflake Bentley."

That night in the library, his profession seemed the most noble occupation imaginable. All that I wanted to do was finish college as quickly as possible, and go out to produce something as moving and beautiful.

Though the solstice is still three weeks away, my flashback has convinced me to use today's snowfall as the first day of winter. There is some precedent for doing so. Winter ecologists have long considered the accumulation of the first twenty centimeters of snow as the beginning of winter. That is when nature is truly divided between the world above and the world below, for the snowpack provides salvation to as many subnivean plants and animals as it proves a bane to those who must bear its weight and scratch through its crusty surface for food.

Even now, the sun is altering the snow, destroying its feathery crystal arms. Within hours, the snow will be transformed into interlocking grains of ice. In the process, it will get stronger, able to hold its shape as it curves off the lips of rocks, trees, and rooftops. Instead of falling to the ground, long strands of snow will loop from fencepost to fencepost, looking like thick white cables.

The report of a breaking branch shatters the early morning silence. The weight of the snow has been too much for another conifer branch. Several

Conifers grow slower but enjoy a longer growing season because they retain their needles and continue to grow in winter.

Bridge over the River Pemigewasset.

have already collapsed, leaving gaping holes in the forest canopy. They have also opened new arenas for competition. Sunlight will now pour through the gaps to the forest floor, and deciduous trees will be quick to respond.

Sunlight is the energy that flows through this system like money through an economy. As in an economy, thousands of communities compete for this resource. In a forest, the two main contenders are coniferous and deciduous trees. They have strikingly different strategies with which to compete. Deciduous trees lose their leaves in winter, but grow faster and more efficiently in summer. Conifers grow more slowly, but enjoy a longer growing season because they retain their needles and continue to grow in winter.

The competition, so intense to the participants, presents us with a striking palette of mixed colors. Each patch of red, each daub of yellow splashed so seemingly randomly on a canvas of conifer green, actually mark specific ongoing battles. The reds are sugar maples holding onto ancient logging roads, the yellows are birches, remnants of past storms. These colorful deciduous trees remind me of my body's own immune system, a system of opportunistic cells that invade a wound and heal it with a scar, a colorful scar that will remain for fifty years.

Down the Notch is an example of another wound that will take far longer to heal. Last February, a multi-ton boulder, dislodged by melt water, broke loose and hurtled down the side of Cannon Mountain, stopping just short of a hiking trail. In its wake, it left a path of snapped and mangled trees.

The boulder is a reminder that these mountains are still eroding at a barely perceptible rate. Usually the pace proceeds pebble by pebble, or frost heave by frost heave—as the frost raises a boulder in winter, then settles it a centimeter further down the slope as it melts in the spring. But sometimes erosion occurs more rapidly. In 1845, an August rainstorm loosened the toe of nearby Crawford Notch, and thirty rock slides clattered down its steep sides, equaling hundreds of years of normal erosion.

The storm was also the event that triggered New Hampshire's worst mountain tragedy. The Willey family maintained the Willey House in Crawford Notch. When one of the rock slides clattered down the side of the Notch, it encountered a huge boulder. The boulder split the rock slide in two, allowing two streams to clatter down either side of the Willey House before rejoining. The story would have had the perfect theological ending if the unfortunate family had not run out of the house to the barn, which was swept away in the cascading tumult of rocks and boulders. Nature has a way of mocking with its disregard of our most cherished beliefs. Hawthorne dealt with this lore in "The Ambitious Guest," his chilling short story based on the tragedy.

Nature also works with a time scale inconceivable to man. Most of us have little real feeling for time before what our parents and grandparents can remember. Perhaps we can imagine Athens, Rome, even Chichen Itza, or our cave-dwelling ancestors. But can we really conceive of the world's age?

During the last two thousand years of history, these mountains have lost only a few feet off their summits. During the last million of human existence, the continental plates have only moved thirty-five feet, hardly enough to register on an atlas of the world. Yet we know that these mountains were formed when the continents were thousands of miles south of here, when the area was underwater, when the granite ledge below me was hot, pasty magma several kilometers below the overlying rock. Such things happen in deep time, not in the thin veneer of time that we inhabit.

Beneath the partially frozen surface of Echo Lake, pipes are ingesting millions of gallons of water and regurgitating them on Cannon Mountain.

Plumes of artificial snow.

A glimmer of sun awakens me from my reveries. Shafts of light shoot from beneath the overhanging clouds. The sun shines on Interstate 93, which snakes its way through these mountains along the path of the ancient retreating glacier. It illuminates a plume of artificial snow now falling on the ski slopes of Cannon Mountain. Beneath the partially frozen surface of Echo Lake, massive pipes are ingesting millions of gallons of water and regurgitating them as snow on the slopes of Cannon Mountain. They remind me that humans have become a force of nature as powerful as the geophysical forces that initiated the Ice Ages, built these mountains, and will eventually wear them away.

I am looking at a landscape that has been shaped by half a billion years of geology, hundreds of thousands of years of glaciation, thousands of years of forest growth, and centuries of human use. With the sunlight glinting under gray clouds on new-fallen snow weighing down the boughs of trees, even the artificial plumes of snow-making machines are a fascinating and beautiful sight, a fitting day to start winter in the White Mountains of New Hampshire.

The coldest air rolls down the mountains and pools in the valleys. © Mount Washington Observatory Photo.

Chapter 5

Coal Hill

JANUARY 2, 1999

"Blam!" Nothing is quite so disturbing to a good night's sleep as the sound of pipes bursting. I look at my watch . . . 3:00 a.m. Too late to do anything about it. Too early to get up. Tomorrow the pipes will thaw. Dirty water will squirt through the wall, pour down the stairs, and puddle up in the basement. Rugs will be soaked, floors will buckle, the ceiling will have to be fixed. I look at my watch again. 3:20 a.m. No one to blame but myself. Why did I try to save a few pennies? Why did I set the thermostat so low?

But wait! Can't I blame the weather? Can't the climate share some culpability? Wasn't December the hottest December on record? Hadn't it almost reached eighty degrees in Boston? Hadn't polar bears been thwarted from starting their winter migrations because Hudson Bay had yet to freeze? Wasn't 1998 the hottest year on record, with Texas posting sixty days with over one hundred degree temperatures? Weren't the nineties the hottest decade on record, with the eighties right behind? In fact, the world has not seen such a rapid rise in temperature since the ice caps melted twelve thousand years ago, perhaps since the ice ages that preceded the dinosaurs.

Wasn't all that enough to justify turning down the thermostat? Of course not. Hadn't El Niño switched to the capricious La Niña, who blew the jet stream south, blasting my defenseless little pipes with cold arctic air.

Was I taking this just a little bit too personally? Was I looking for someone to blame other than myself? Of course! Humans have been doing that for generations. First it was mischievous weather goddesses, then malevolent Russians, now it's global warming. Will we ever do something about it? Perhaps when the world's average annual temperature tops sixty degrees in the next decade or so?

Probably not. Scientists estimate that the world would have to cut back on the use of fuel by 60 percent in order to reverse global warming. That would mean that the entire population of the world would have to start sharing the amount of oil that the United States used prior to World War II. That means we would have to throw out our PCs, sell our SUVs, drive 60 percent less, travel 60 percent less, and give up two-thirds of our jobs and incomes. It might happen, but it will probably have more to do with the world's running out of oil than voluntary compliance.

The human brain is notoriously poor at dealing with such long-term problems. It much prefers to obsess about short-term problems such as wars, burst pipes, Y2K bugs, and that endlessly fascinating topic, the sex lives of our leaders. Blame it on our primate ancestors if you must. Evolution made us this way.

Okay, so we're going to be stuck with global warming for most of the next millennium. What should we do about it? We might as well start by looking at what it will do to individual ecosystems.

So where's the best place to look? Right in our own backyards. My backyard happens to be Coal Hill, a small mountain behind our house. My Socratic monologue has decided my fate. I will get up, explore the hill, and forget about the damn pipes.

After a quick cup of coffee, I strap on my skis and head downhill. The ground is covered with five inches of new-fallen snow. The woods appear static and natural, the forest primeval, unchanged since the retreat of the glaciers—but they are something else.

My first stop is the Gale River at the bottom of the hill. It is dank and clammy place. Clouds of mist envelop the stands of black spruce that loom above the river. A foggy stillness hangs in the quiet trees. At night, the coldest air oozes down the mountain and pools in this valley. Only black spruce can thrive in the standing water and nutrient-poor soil. They can even reproduce vegetatively by sprouting roots from branches that touch the mossy, water-saturated soil. I can see a few of the branches already turning upward toward the light as small, independent, new trees.

If this were a straightforward conifer forest, hiking up this mountain would be like walking through a well-ordered ecological experiment. For every thousand feet I climbed, the temperature would drop three degrees and the rainfall would increase eight inches. Every thousand vertical feet would be like driving north three hundred miles. White spruce would

A textbook example of succession: black spruce mopheads and krumholz (top), *conifers* (middle), *hardwoods* (bottom).

Coal Hill is nothing like the textbook example of natural succession I was hoping for. A beech tree reaches for sunlight.

replace black spruce as drainage improved and nutrients became more available. Near the top of the hill, red spruce and balsam fir would replace the white spruce.

At the timberline, the black spruce would again predominate. Only they can tolerate the nutrient-poor soil and cold temperatures above four thousand feet. The versatile trees have an added advantage that allows them to grow above the timberline. Instead of growing up, the trees grow sideways, sprouting adventitious roots from their branches. This allows them to form large mats of low-lying vegetation called krummholz or "crooked wood." During winter, snow covers the krummholz, protecting the branches from wind-driven ice and cold.

But Coal Hill is nothing like the textbook example of natural conifer succession that I was hoping for. First, it is a mixed northern hardwoods forest with about equal numbers of hardwoods and conifers. Second, it has strange anomalies. A gnarled old apple tree grows in the shade of a copse of spruce. The spruce are surrounded by rotting logs of birch, a fast-growing species of tree that needs light. Further up the mountain, broad swaths of yellow birch and maple grow. Beech trees thrive along the drier ridges.

I'm confused. What should be a short walk through a well-ordered example of natural succession is morphing into a treacherous maze of trick questions. Is this the ecological exam from hell, or just the exam of the ecologist from hell? What's going on? Where is the answer sheet?

Two centuries ago, this part of New Hampshire supported agriculture and light manufacturing. Instead of forested mountains, these were open

The effects of sea-level rise.

fields and orchards. The Gale River had been dammed to power saw mills and iron works.

By the 1830s, only 10 percent of New Hampshire's original forests remained; 90 percent of the state was cleared land. There was no need for today's ubiquitous moose- and deer-crossing signs. So little forest cover remained that moose, deer, and wild turkey were all but extinct. The bear population was down to less than fifty animals.

What happened? Almost half the population of many of New England's small agricultural towns moved west. The Midwest's broad tracts of rich farmland were ideally suited to the newly invented MacCormack Reaper, and new railroads made it possible to ship produce back to the East Coast markets. By 1860, the forests were starting to reclaim thousands of abandoned farms, fields, and pastures.

Today, New Hampshire is 90 percent wooded and 10 percent open, exactly the reverse of the situation two hundred years ago. The results are impressive. Deer, moose, bear, turkey have made an impressive comeback, although coyotes have displaced wolves as the top predators in the food chain. Northern New England is probably the largest area of the world ever to have undergone such massive revegetation after human development.

But the result is also unsettling. All I can see are the merest traces of humankind's past activities: an apple tree here, an abandoned field there, stands of deciduous trees where the skid rows of loggers once scored the mountains.

It will be difficult to discern the subtle changes caused by global warming within this confusing mix of humanmade and natural changes. In other areas of the world, the changes are more obvious. Along the coast, oceanfront homes are being swept out to sea because of rising sea levels and more frequent storms. In Africa, the desert is encroaching on the former agricultural land of the Sahel. Here in New England, the effects of global warming will be subtle, and often counterintuitive.

For instance, milder winters may actually be more damaging to plants and animals than harsh winters. This is because the temperature will tend to fluctuate back and forth across the freezing point. The effect is a little like my water pipes. The real damage comes not so much when my pipes freeze but afterwards when they thaw. The same will be true for nature.

Northern plants and animals can adapt effectively to cold weather by changing physiologically and seeking the protection of snow. So it is actually easier for them when it freezes, snows, and remains that way until spring. It is more difficult when they gain, then lose the protection of snow and have to readapt to weather above and below freezing. We see common examples of this when bulbs sprout in a winter thaw, then die when it turns cold again. Even rocks do poorly when temperatures cross and recross the freezing point. They crack and erode more quickly as water freezes and thaws in their fissures.

Global warming may also affect the types of storms we will have in New England. Today is close to the anniversary of a the 1998 ice storm that paralyzed much of southern Canada, New York, and New England. From January 7 to January 11, freezing rain splattered to the ground, then froze on contact. Streets, trees, and power lines were glazed with a sparkling coat of beautiful but deadly ice.

The storm toppled a thousand transmission towers, three hundred thousand utility poles, and untold billions of trees. Over three million people lost their electricity and water. A month later, thousands of them were still huddled around their fireplaces eating cold meals by candlelight. They survived, but may very well bore their grandchildren to death by the retelling of their tales. We New Englanders do so much like to talk about the weather.

The ice storm was rated as the most damaging storm in New England's history. Of course, the year closed with Hurricane Mitch, history's deadliest hurricane. Scientists have predicted that we will get more of this kind of extreme weather as global warming increases.

But the prime reason that the 1998 ice storm was so destructive was that the temperature was hovering near thirty-two degrees. The air temperature was warm enough for rain to fall, but the ground was so cold that the rain froze when it came in contact with anything solid. We can expect this to happen more frequently as global warming brings milder winters.

So some of these broad swaths of downed trees can be considered to be the indirect result of global warming. The scars will remain on the mountainsides for generations to come. But won't they eventually be eradicated by natural succession? Maybe not.

Fast-growing pioneer deciduous trees will grow into the open spaces, mature, and die. But in fifty years, when conifer trees should start to overgrow the pioneering deciduous trees, will conditions be the same or will they have tipped in the hardwoods' favor? Will places like Coal Hill become more solidly hardwood forests than conifer forests?

It has happened before. In New Mexico, a drought pushed a ponderosa pine forest back two kilometers in less than five years. The ponderosas were

The ice storm of 1998 will affect forest succession for decades. Courtesy *The Courier* newspaper.

replaced by a mix of piñon and junipers. The drought ended decades ago, but the ponderosa have not grown back to their former limits. Global warming has tipped the balance in favor of the faster-growing, weedier species.

Such ponderings have brought me to the top of Coal Hill Road. It is a sad and lonely place. All I can hear is the mournful sighing of trees as they creak and sway in the violent wind. They are all the same age, all the same height. It feels like they have witnessed some great tragedy. In a sense they have.

Two centuries ago, Coal Hill was the place where people made charcoal for the Franconia Iron Works. The top of the hill was entirely denuded, and large, covered sheds of wood smoldered for months on end. The smoke from the conflagration crept down the valley and hung like a pall over the town. The practice came to an end when the Franconia Iron Works went out of business in 1872. They couldn't compete against iron works in Pennsylvania that were close to coal, a far more efficient, cleaner fuel.

Today, all traces of the charcoal graveyards have disappeared. Only the towering pine trees remain. Only they could sprout and grow fast enough to beat out the grasses in the ashy soil. But these trees couldn't be that old.

There must be more to this story than meets the eye. I will have to come back up in spring to figure out the trees' full story. For now, these pine trees make a fitting memorial to a place that witnessed so much destruction.

Beside the pines is a large field overgrown with pin cherries, shrubs, and a few small maples. But the end of each twig has been nibbled off. I try one. It is not an exciting cuisine. The pin cherries taste slightly nutty, the maples like nothing at all.

A small herd of white-tailed deer browsed here, then lay down to ruminate. This winter has been difficult for them. I can see where the snow crust has lacerated their ankles and forced them to adopt a bounding, energy-consuming gait. They have trampled out this yard to help conserve precious energy. They have little to spare. The twigs supply the deer with less energy than they expend in browsing for the tasteless morsels. The deer are already living off the fat they built up in autumn.

On the edge of the field, I see a splotch of blood in the snow. It is below a maple tree where a buck had rubbed his antlers, probably trying to remove the now useless appendages.

It feels like these trees have witnessed a great tragedy. Courtesy *The Courier* newspaper.

Pine trees dwarf Coal Hill skier.

Winters are especially hard on old bucks. During the fall, when they should be eating to accumulate fat, they have to expend valuable time and energy courting does, fighting off competitors, and maintaining their harems. It is the price male deer play for their polygynous behavior. Their genes will be passed on, but the individual deer will suffer.

I follow the old buck's tracks. He was being followed by a dozen coyotes. Perhaps they were drawn by the smell of his blood, perhaps they sensed he was vulnerable from the accumulated energy deficits of too many difficult winters. Whatever the reason, the coyotes seemed to have been tracking him for several days.

Deer kill.

But last night it happened. The pack surrounded the buck, hectoring and biting at his flanks. I could see where he fought back valiantly. His hooves were sharp and his antlers deadly. Some of the blood belonged to his enemies. Yet each time the buck lowered his head, he exposed a flank and expended more energy. He was growing tired from lack of sleep and too little food.

As several coyotes nipped at his heels, one lunged for his neck and severed an artery. More blood gushed onto the trampled snow. The coyotes moved closer. Now the old buck was on his knees. Two coyotes leapt on his back and another grasped his neck and would not let go.

Perhaps the fight no longer seemed important to the deer, perhaps he was just too tired to continue. Whatever the reason, the result was the same: tracks, snow, a piece of bloody backbone, a well-gnawed leg. The rest of the carcass has been scattered through the forest.

Aftermath of a blizzard. © Mount Washington Observatory Photo.

Chapter 6

A Mid-Winter Blizzard

JANUARY 16, 1999

It is frigid on the top of Cannon Mountain. I am standing in a small metallic cage suspended from a thin cable hundreds of feet above the ground. It sways precariously in the gale-force winds. This is the early morning inspection run of Cannon's aerial tramway. The ski patrol is going to decide whether it is safe to open the mountain, safe to operate the tram.

It does not look good. I can't see Mount Lafayette. I can't see the ground. I can't see the sky. All I can see is swirling white snow. I feel like I'm in one of those toy snow dioramas that has just been given a vigorous shake. I peer into the whiteness. I see nothing but the occasional top of a frosted tree fifty feet below. We continue up the mountain.

I am searching for the tower that marks the midpoint of our ascent. If the tram gets stuck, at least I can climb down to safety. I've heard those stories about impatient skiers who jumped out of stalled tramcars and broke their legs. I can imagine the last thoughts of the Italian skiers who plunged to their deaths when a hot-dogging, low-flying marine jet severed the cable of their gondola. At least today is too snowy for planes to be flying.

At last, the tramcar rises as we approach the tower. The empty car of the descending tram trundles past, then slips quietly back into the swirling snow. Now it is silent. We are alone once more. Our car drops several feet and starts to sway more violently. We are at the most perilous point in the ascent. If the swaying gets worse, the operators will have to stop the tram and leave us dangling in the air.

I look at the trapdoor in the ceiling of the car. It is not an altogether reassuring sight. What are we supposed to do, squeeze through the tiny opening, clamber over the icy roof, grab the cable and brachiate, hand over greasy hand, to the nearest tower hundreds of yards away? What if I get stuck halfway through the trapdoor with a bunch of frostbitten ski patrol people impatiently pushing me from below? Am I feeling the effects of thin air? Is this what Jon Krakauer experienced on Mount Everest?

Finally, the welcoming maw of the summit station looms into view. The operator waits for the swaying to stop before jostling the car between two steel bumpers. The head of the ski patrol opens the door. A blast of cold air takes our breath away. I can feel itchy little crystals of ice forming in my nose.

But this is not the worst weather on earth. That distinction is reserved for neighboring Mount Washington, which can have hurricane-force winds and freezing temperatures any month of the year.

In 1934, Mount Washington meteorologist Salvatore Pagliuca tied a rope around his body, forced open the door of the observatory, crawled across the deck, and checked the weather instruments while his fellow weatherpeople held him down. He was able to record the weather without being blown sideways off the mountain. The wind speed was 231 miles per hour, the lowest temperature minus 46 degrees Fahrenheit. William Lowell Putnam wrote rather archly of Mount Washington's record, "There may be worse weather, from time to time, at some forbidding spot on the Planet Earth, but it has not been reliably recorded."

Mount Washington also used to hold the dubious distinction of being North America's most murderous mountain. But that was before ill-prepared socialites started spending thousands of dollars to be pushed, pulled, and sometimes carried to the top of such exotic locales as Danali and Mount Everest.

When David Braeshears made a film about the deaths of climbers on Mount Everest in 1997, he shot his most dramatic scenes later on the windblown summit of Mount Washington. When summertime drivers reach the top of Mount Washington, the first thing they see is a large board that records the names of the two hundred-plus climbers who have frozen to death on the deceptive mountain.

Certainly the people who died on Mount Everest could have done so a lot less expensively on Mount Washington. But I suppose it would have lacked the requisite cachet. They would have had to share the honor with cars—tacky cars sporting those big bumper stickers that say, "This car climbed Mount Washington!" Where's the social cachet in that?

So I have opted for something less than the worst weather on earth. But why is northern New England home to such horrendous weather?

First, two stationary highs develop and hover over Hudson Bay and Bermuda. They funnel storms from the West Coast, the Midwest, the Gulf Coast, and the Atlantic Seaboard so that the storms converge on northern New England.

Second, the Canadian high spins clockwise so it pulls cold arctic air south. The low pressure storms spin counterclockwise, so they also push arctic and Atlantic air into northern New England. Winds from the storms blow up the flanks of the White Mountains, often meeting vastly different air masses on the other side. The air temperature plummets as the wind ascends, and New Englanders get buried under several more feet of snow, sleet, rain, or hail.

Today's storm is an example. It started last night when cold arctic air from Canada collided with warm, moisture-laden air coming off the Atlantic. The air masses converged over Mount Lafayette, striking its flank several times with lightning. Our house shook with the force of the wind. This morning the snow continues unabated.

I take a few tentative steps outside. I can already feel the muscles tightening in my neck and shoulders. This is our body's way of generating heat. The next step is shivering, then hypothermia, then you die.

My hands are already cold. I take off my gloves to blow on my fingers. My brain's hypothalamus has signaled my pituitary to squirt noradrenaline into my blood. This has caused my veins to constrict. The constriction has shunted warm blood into the deeper veins that protect my brain and heart. This is the body's way of preventing heat loss. As the veins lose their power to react to noradrenaline, the veins will dilate and blood will surge back into my fingers, causing them to tingle.

In some people, such vasodilation happens sooner. My father could fish for hours barehanded in freezing water. Perhaps it was winter training in the Tenth Mountain troops that had conditioned his hypothalamus to keep his fingers warm and nimble during cold weather. Some people have that ability, but even their cold tolerance can't compare with animals. Humans are essentially tropical beings, who must use technology to protect ourselves from the cold. . . . Perhaps I can use that as my excuse when someone asks why I rode the tram up Cannon rather than hiked to the summit of Mount Washington.

No, this is cold enough for me. The trees are windblown and stunted.

Rime ice, Artist's Bluff.

Bits and pieces of twigs and needles litter the trodden snow. Even though Cannon Mountain is almost fifty feet below the timberline, you can see the effects of the high altitude.

The spruce trees look like weather vanes. A delicate latticework of rime ice faces the prevailing wind. The rime ice forms when water-saturated air is super-cooled as it blows over the top of Cannon Mountain. The air becomes a cloud of fog that freezes on everything it touches.

The ice is beautiful but deadly. When the wind changes direction, it blows off the tree's ice-encrusted foliage. Some trees have already lost a fifth of their vegetation. Next summer, they will spend several months regrowing their photosynthetic capacity. Many may never recover.

Rime ice, Mount Washington.
© Mount Washington Observatory Photo.

Some of the trees have a curious mophead appearance. They are being sculpted by the same wind that I can feel lashing at my legs. The wind carries tiny, sharp crystals of ice that can strip off foliage and lacerate the bark of small trees. But the wind is strongest a few feet above the surface. Above that, it is calmer, so twigs, foliage, and buds can survive. Over many seasons, the trees have built up these comical mopheads of vegetation above the zone of greatest abrasion. It gives me an inkling of the role that wind, ice, and water play in creating the timberline.

But, as cold as it is, this weather is nothing compared to the last ice age. During that time, Cannon was sometimes above, sometimes below, the half-mile thickness of ice that scoured this valley. But the last ice age was a mild ice age. It came, it went, and it will undoubtedly return. However, if we travel back into deepest geological time, a time when New Hampshire was underwater somewhere near the equator, our world experienced an ice age that almost brought life to a premature end.

Scientists have only recently discovered the evidence for the cataclysmic Verangerian Ice Age. It started 760 million years ago, when the earth's cli-

Rime ice on slender pole.

© Mount Washington Observatory Photo.

mate was still warm. Nearly all the continents on the planet had clumped together to form the supercontinent of Rodinia. The land was still devoid of life, but it was a different matter at sea. Rodinia had started to rift apart, creating broad expanses of shallow seas replete with simple microbes and algae, the only organisms yet evolved.

The primitive creatures made up in numbers for what they lacked in sophistication. Life was in overdrive, photosynthesis rampant. The simple life forms drew billions of tons of carbon dioxide out of the atmosphere and incorporated it into their bodies. Sediments eroding from the newly formed mountains buried the organisms on the sea floor, sequestering the greenhouse gas from the atmosphere. The result? The world started to cool.

By this time, Rodinia had rafted south to straddle the South Pole. This was a dangerous situation. Now there was a huge landmass on which snow and glaciers could accumulate, far larger than the present continent of Antarctica, whose small size acts as a brake to limit the extent of southern glaciation. The glaciers cooled the planet further by radiating heat back into space.

Eventually, the entire supercontinent was covered with glaciers, and the oceans were frozen solid with pack ice. Our world looked like a hard-packed ice ball, the kind you might mold all day to bean a particularly nasty neighbor. There were no patches of blue ocean, no clouds in the sky, no green on land. The entire planet was a smooth expanse of ice, with only a few patches of bare rocks protruding to mark the highest mountains.

Oceanic ice blocked the sunlight, shut down photosynthesis, and killed off life almost before it had begun. Among the only organisms to survive were chemosynthetic bacteria, which continued to live off the earth's energy in the scalding vents of underwater volcanoes. Even snow stopped falling because no open ocean remained to supply water vapor to the atmosphere.

Then the earth stayed that way for ten million years, ten times longer than humans have been on earth. The sun was no use. It was 7 percent dimmer than today, and its energy was being radiated back into space. It seemed as though our planet had been short-circuited, destined to remain a dirty, lifeless snowball drifting through space like one of the frozen moons of Jupiter.

So what finally happened? Our planet saved itself. As plate tectonics progressed, volcanoes started to erupt underwater. They spewed out massive plumes of hot water and belched billions of tons of carbon dioxide back into the atmosphere. The climate started to warm, the polar ice caps disappeared, and the planet was in danger of returning to a runaway greenhouse world.

What happened then, you ask? Life came to the rescue once more. Millions of new species evolved to draw down the carbon dioxide budget and cool the planet. We were back in the icehouse! The earth swung back and forth this way, five times during 250 million years. Like a thermostat honing in on a setpoint, plate tectonics and evolution worked hand in hand to bring our planet back into equilibrium—an unconscious yet marvelous partnership. Some would call it the invisible hand of Gaia.

Cannon Mountain ski patrol.

Finally, the earth's temperature settled down at a level several times hotter than now. Carbon dioxide levels were 350 times higher than at present. Evolution went into overdrive. We had the Cambrian explosion, which led to virtually all the basic life forms we see today.

But it had been a close call. There were no guarantees that our world would recover. Our neighboring planets Mars and Venus never did. One was trapped as an icehouse world, the other as a runaway greenhouse world. We were just lucky, and had time, a lot of time, on our side. It makes a blizzard on Cannon Mountain seem not so bad after all.

And perhaps the force is still with us. The ski patrol has decided it is too cold to open up the mountain. I can take the tramcar to the bottom, stoke up the fire, and ride out the storm with a good book.

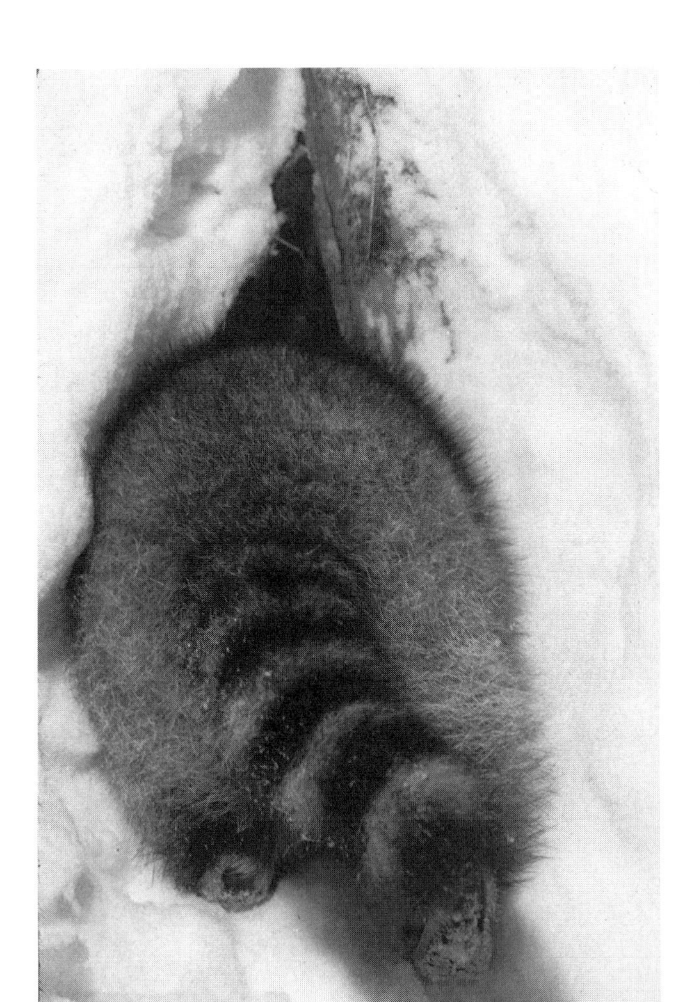

All I want to do is burrow back under the covers; raccoon entering burrow. Ted Levin.

Chapter 7

Valentine's Day

Blame It on the Oxytocin

FEBRUARY 14, 1999

It is warm and cozy beneath the blankets. My wife snuggles beside me, her body exuding a potent cocktail of warm, rich, mammalian smells: pheromones, body odors, endocrines, and ectocrines. I poke my head out of the covers—another cold day. All I want to do is burrow back under the blan-

Snow patterns. Courtesy *The Courier* newspaper.

kets to reinforce the bond we forged last night. What do I have to do? Go outside and make the donuts.

Coffee helps purge me of the warm, fuzzy hormones. My body is now primed with more businesslike neurotransmitters. I am alert and ready to forage.

Outside, the meadow lies under an icy sheen of crusted snow. Only the tall, dark stalks of weeds jut through the white mantle. The forked stems of mullein look like ornate candelabras; the dry, brown stalks of fern look so different from their summer fronds. The translucent leaves of mustard plants rattle dryly in the winter wind, and the swollen galls on goldenrod offer the only evidence of life. Later, when the sun shines, tiny primitive snow fleas will emerge from the leaf litter to pepper the crust of the melting snow.

A scolding chatter announces that a red squirrel is annoyed with my presence. He is more than usually aggressive, as his mating season is fast approaching. He guards a neat pile of scales that he has husked off a spruce cone in search of its seeds.

There is other evidence of his work. The snow is littered with the green tips of spruce twigs. The squirrel has nipped them off while foraging for cones. Now they make easy pickings for deer that are beginning to feel the effects of winter. It takes valuable energy for the deer to dig beneath the snow for food, and by mid-winter they have already browsed the needles off the lower branches of the trees that they can reach. In harsh winters, deer seek out such litter piles to help provide the margin of error they need to survive.

Sometimes a bobcat will hide in the branches above such a pile of litter. When a deer comes to feed, the bobcat drops on the deer's back and bites through his spinal cord. The deer throws back his head and usually dies in that position.

Yesterday I surprised a bobcat crossing the snow-covered road. She leaped back into the woods, looped around, and crossed the road after I had passed. She has an active hole near here. We have often smelled her scent while walking down the road. It smells like a cat box. It's hard to think of her killing a deer, but there is blood in one of her paw tracks. Perhaps it was just from a snowshoe hare. Their huge tracks are also clearly visible.

The red squirrel pauses to scold me once more before dashing into his subnivean tunnel. The snow has made life much easier for the small mammals. Heat rising from the earth has melted the bottom of the snowpack, creating pockets lined with stalactites of hoar ice.

The rodents can now tunnel beneath the snow safe from the eyes of weasel, fox, hawk, and owl. Occasionally, you see where a rodent has run across the surface of snow before plunging back into its next set of tunnels. If the tracks pause to walk or trot, they are usually the work of voles or shrew; if they gallop straight to safety, they are the tracks of mice.

Beneath the snow, mice are preparing to do what mice do best—reproduce. It's fortunate too, because all these small rodents are the prime intermediaries between the plants at the bottom of the food chain and the predators at the top.

But they are not reproductive automatons. I like to think of them living beneath the snow enjoying the advantages that enabled them to survive as the dinosaurs succumbed. First, they were warm-blooded and had fur.

Snowshoe hare burrow.

Monsters make better exhibits than cat-sized predators. Harvard Museum of Natural History.

These attributes allowed them to be active at night when dinosaurs were asleep. Dinosaurs needed sunlight to stay warm.

Second, early mammals were small. People used to think that their smaller size helped mammals avoid the notice of dinosaurs. But that is probably an artifact of our museums. Displays of monster dinosaurs make much better exhibits than fossils of cat-sized predators. Just as many more cat-sized predators than polar bear–sized predators exist today, during the age of dinosaurs there were many more Ornitholestes-sized dinosaurs than *Tyrannosaurus rex*–sized dinosaurs. It's just that the monsters left better fossils and make more impressive exhibits. Who would go to a Jurrassic Park full of puny but safe little Ornitholestes?

But early mammals' small size really helped them when an asteroid the size of Texas slammed into the Gulf of Mexico. The asteroid kicked up clouds of dust that obliterated the sun and brought photosynthesis to a halt. Dinosaurs started starving to death. But mammals, being smaller, could survive on the few leaves and seeds still available—to say nothing of the huge mounds of dinosaur flesh rotting in the gloom.

But mammals had an even more potent advantage: sex, breasts, and the rich mix of neurotransmitters that combined the two. Together they led to the large, socially intelligent brains we hold so dearly today.

First, let's look at breasts. Lactation allowed mammals to forage for food while their young were safely ensconced in their burrows. When mother mammals got home, they could feed their infants milk; white gold alchemized from grass, seeds, and whatever else the mothers had come across during their nocturnal travels.

Milk allowed mammals to nourish their young until they were wily little sub-adults able to fend for themselves. Compare them to dinosaurs who had to hatch out of eggs, find food, and defend themselves when they were as small and defenseless as a baby lizard. No wonder dinosaurs succumbed to the asteroidal holocaust.

But female mammals did not evolve just to be passive receptacles for sperm and fetuses. They are the original pro-choice animals. If food is in short supply, a mother mouse can reabsorb her developing embryo. If her mate has died she can abort her fetus. This will allow her to find a new mate and start over again with a better chance of success. Some female mice will even kill off their offspring if a strange male has supplanted the mate and has infanticidal intentions. This will allow her to mate with the new male and successfully raise a litter. What can we make of such maternal behavior? Is it also present in higher mammals?

Not far away from where I stand, steam rises from the snow. This is the lair of New England's largest mammal. Right now her den is a warm, broody chamber. A week ago, the female black bear hardly awoke to a new sensation. It was the birth of her two naked cubs. It was an enviable birth; the cubs only weighed eight ounces, little more than a stick of butter. The chipmunk-sized cubs smelled their way toward her nipples and remain there gaining weight while their eyes have time to open and their ears time to unfold. Their mother sleeps blissfully through it all.

Bears are also prototypically pro-choice mammals. If the mother bear had not found enough fatty food, beech nuts, and acorns this past fall, her cub's blastocysts would not have implanted on the wall of her uterus. She would have skipped having cubs, waiting for a more propitious season. If she can't find enough food, a mother bear will sometimes abandon a single cub. That way, she can mate again, and take the chance of raising two cubs or even three cubs during the precious time it would have taken her to raise the singleton. Anthropologists have even found groups of women on high-fat diets who have evolved a genetic predisposition to have twins, rather than single babies.

How do female mammals make these intricate calculations? How does a female mouse know to reabsorb her embryo when no male is around to help care for her offspring? How does a mother bear decide to abandon a single cub, on the chance that she can raise two in the future? The answer lies within their brains and in that rich cocktail of mammalian hormones that guide their lives.

First, there is prolactin, one of those ancient all-purpose hormones. Amphibians used prolactin to induce metamorphosis long before evolution ever conceived of mammals. Mammals use prolactin to control lactation. After birth, a rise in blood-born prolactin signals the onset of milk production. As an infant suckles, the nipple signals the brain's hypothalamus to excrete more prolactin. As long as the infant keeps breast feeding, prolactin is also the hormone that prevents the mother from resuming ovulation and having other resource-consuming infants. Talk about sibling rivalry!

Oxytocin is that warm, fuzzy hormone that makes being a mammal so much more enjoyable than being a frog. It is a purely mammalian invention that is manufactured in the brain, ovaries, and testes.

During pregnancy, a mother's brain changes. Her pituitary grows larger and her uterus develops receptors for oxytocin to help induce powerful labor contractions.

But oxytocin is also a bonding hormone. During suckling, oxytocin passes from the mother's milk to her infant, giving them both a sense of blissful, almost sexual satisfaction. This is the time when human mothers "fall in love" with their babies and when husbands have been known to feel neglected and just a little bit jealous.

The brain-breast circuit can become so conditioned that the cry of a hungry infant is enough to trigger "let down." The brain secretes oxytocin, which signals milk to flow toward the nipple. A business blouse may be moistened; but not to worry, it is just a gentle reminder that we are not so far removed from our mammalian cousins as our surroundings might suggest.

Oxytocin is also released when the back is massaged or the breasts or testicles are manipulated. This helps people bond when they are making love. It also causes afterglow, that warm, free-floating bliss that follows orgasm.

During birth, oxytocin and its artificial cousin pitocin induce heavy labor. When you hear, "Hit 'er with the Pit," you know the baby will soon be on its way—and that the obstetrician is probably late for his tennis match.

But oxytocin is also responsible for medicine's best kept secret: that the birthing area is the most wildly romantic room in the hospital. As the midwife massages the mother's back, oxytocin is contracting the mother's uterus and suffusing her brain. The mother is falling in love with the obstetrician, the father is getting turned on by the anesthesiologist, the anesthesiologist is eyeing the nurse. The baby is born just in time so that everybody can hug,

How does a mother decide which fox pup to abandon?
Mass. Wildlife/Bill Byrne.

Alpenglow; an altogether different kind of afterglow.

kiss, and return to their daily lives, only slightly baffled by the rush of confusing emotions and intense bonding they have just experienced. Postpartum blues may just be the result of withdrawal from the surge of this natural potent opiate.

In old age, oxytocin is the hormone of peace and bonding. It is there when couples hold hands, groom, and perhaps croon, "When the children are asleep I'll dream with you, about the times that we had, and be glad, that it all came true . . ."

But I must be going. My wife waits at home, sitting at the very top of our mammalian order; oxytocinally primed to protect, love, and lactate. I think I will go inside to enjoy the glow of the fire. Who knows, I might even get to enjoy the afterglow.

Bobcat at night.
Mass. Wildlife/Bill Byrne.

Chapter 8

Deep Time

> *The odds against a universe like ours emerging from something like the Big Bang are enormous.* —Stephen Hawking, *Boston Globe*, 1997

MARCH 6, 1999

It is a moonless night in March. My dog snuffles through the new-fallen snow. He has picked up the scent of a passing creature. I switch on my flashlight to see a bobcat disappearing into the understory.

There is a pungent musky odor in the air tonight. Both the dog and I sense that another being is with us. The beam of the flashlight sweeps through the forest. It falls on the large, hairy flank of a sleeping moose. She stands and gives us a cold, hard, significant stare. We share a moment of mutual indecision. Finally, she turns and trots noiselessly through the deep snow. In summer, she would have looked heat-stressed and shaggy, comic if not pitiable. Tonight, she looks at least well-adapted to the cold and snow—if not a little noble in her carriage and power.

I switch off the flashlight. The stars emerge with startling clarity. The lights of a high-flying jet navigate silently through the wispy void. The stars are becoming intense, disorienting. Mercury, Venus, and Jupiter overpower the Milky Way. I realize I am no longer standing on earth observing the winter sky. I am the winter sky. This universe is part of me; I, part of it.

A small smudgy cloud hovers to the left of Jupiter. I feel as though I could lift off and travel to it, but it has already traveled two million light-years to me. Its light started coming here when these mountains were still being formed, when the earth was just entering the Ice Age, when humans were still bipedal apes roaming the dry savannas.

The smudge is the Andromeda Galaxy, famous because it forever changed the way we look at our cosmos. In 1924, Edwin Hubbel was sitting in the Mount Wilson Observatory stifling shivers as he scanned the stars of Andromeda. Suddenly, he discovered what he was looking for, a Cephalid variable. He knew he could use the changing brightness of the star to measure the distance of Andromeda from earth.

Overnight, the universe got very big and not very cozy. Hubble had shown that Andromeda had to be a separate galaxy from our own. With the discovery came the realization that not only were we not at the center of the universe, but we were not very special either. We were rather insignificant creatures, on an insignificant planet, on the edge of a rather small galaxy, indistinguishable from billions of others. It was a major blow to our egos.

Earlier, another astronomer, Vesto Melvin Slipher, had taken spectrographs that showed that the Andromeda Galaxy was rushing toward us at three hundred kilometers per second and that other nebulae were rushing away from us at six hundred thousand kilometers per second. Although he was roundly congratulated for his work, nobody really knew what Slipher's redshift observations meant. It was only when Hubble showed how distant other galaxies were from our own, that Slipher's observations made sense. Not only was the universe immense, but it was flying apart on the fading waves of the Big Bang.

I've always had a problem with the Big Bang. It smacks too much of the biblical interpretation of creation, and we all know how well that has held

Beech tree in snow.

up. But the main fault of the Big Bang is that it only gives God one shot at getting it right. As the physicist Stephen Hawking once said, "The odds against a universe like ours emerging from something like the Big Bang are enormous." Energy had to devolve into four forces, matter had to coalesce into stars, stars had to produce heavier elements, supernova had to explode, heavy elements had to incorporated into planets, and planets had to evolve just the right conditions for life. Everything had to happen with split-second timing, in just the right way, in such the right sequence.

Granted, the Big Bang provided fifteen billion years and millions of suitable planets on which to experiment; but there were no guarantees. Our universe could have had entirely different physical properties. Time could have run backwards. Our universe could have had too many or not enough dimensions. Antimatter could have annihilated matter, all matter could have remained as hydrogen or turned entirely into helium. All possibilities were possible and most were far more possible than what actually occurred. Besides, what happened before the Big Bang, and how can you turn nothing into something in the first place?

Now physicists are testing a theory that neatly answers these problems. It is the eternally self-replicating universe devised by the Stanford physicist Andrei Linde. His theory implies that our universe was not made out of nothing, but that space, time, energy, and matter have always existed as different forms of each other. He says that our universe, created by the Big Bang, is just one of an infinite series of universes that have come before and will continue long after ours has petered out.

Suddenly our cosmos has become a lot older, a little less special, and a lot more probable. Instead of a paltry fifteen billion years, we now have an infinite amount of time to create life. Instead of a million chances, we have hundreds of millions of chances to recycle space, time, energy, and matter

Time passes. Courtesy *The Courier* newspaper.

Winter road.

into new universes with slightly different properties. That sounds a lot like evolution, doesn't it? Perhaps there are even gene-like sets of instructions that are passed down from one universe to another to guide their offsprings' development.

We have already seen that people, even life on earth, were not preordained. Millions of things could have gone wrong. Most of them did. We had comet impacts, ice ages, greenhouse conditions, and oxygen pollution. Chance, luck, and coincidence all played a part, but the most noteworthy thing is that once evolution gets started, it is damn hard to stop it. Millions of species may be snuffed out by a single event but evolution just keeps trundling along.

Just as it seems ridiculous to think that all life was created on the first try, it takes an astonishing amount of faith to think that a perfect universe could be created in just one crack. But given an infinite amount of time, it becomes almost unthinkable that evolution could not get it right.

We have gone from thinking of the development of an individual universe, to thinking of the evolution of species of universes. The former is a chancy proposition, the later almost a foregone conclusion. Suddenly God can make millions of mistakes. He has billions of chances to get it right. He doesn't have to be all-knowing, all-powerful. Of course, if you have evolution, do you really need a god in the first place?

However, the remarkable aspect of this theory is that it suggests that a universe is born, grows, reproduces, and dies. That pretty much fits our definition of life. But the other thing that evolution needs is some way of passing on accumulated knowledge, a kind of cosmic DNA, if you will.

Maybe that's where intelligent life fits in. Why are we so obsessed with discovering how our cosmos came to be? Perhaps we are like proteins in a reproductive cell, compelled to discover how our universe unfolds, to encode the information on some form of cosmic DNA, and shove it through a black hole to mingle with DNA from another universe. Is the role of intelligent life to be unwitting participants in the conception of a new, slightly different, hopefully more successful universe? If so, we would be no more aware of our role than are the reproductive cells in our own bodies.

Perhaps even now some particle physicist is in his lab creating just the right conditions in a particle accelerator to set off the next big bang. That would certainly keep the "special role for humanity" folks happy. But who is to say it would be a human particle physicist?

At any rate, we will never know that the next big bang has even occurred. It will simply create its own inflating time and space, forever parallel, yet forever separate from our own.

I find it somehow reassuring to think that I am part of an eternal, evolving universe. I like to think our earth might even have some small, unwitting, reproductive role in that process. I like to imagine that somewhere off amongst these many stars, energy is rearranging itself, a singularity is forming, a new universe will soon be born. It just seems so right on a star-filled night in March.

Hyperactive chickadees. Courtesy *The Courier* newspaper

Chapter 9

A Quest for Green

The jabberwock with eyes of flame,
Came whiffling through the tulgey wood,
And burbled as it came. —Lewis Carroll, "Jabberwocky"

MARCH 16, 1999

Spring does not so much bloom in New Hampshire as does winter release her icy grasp in reluctant fits and starts. Last week we had a raging snow storm, yesterday a cold hard rain. Today it is warm but moody and miserable. I'm seized with a restless yearning to see real signs of spring; the reddening of buds, spring birds at the feeder, the appearance of new green leaves.

Every day I find myself, coffee in hand, checking the bird feeder for new arrivals. But it is only the same flock of hyperactive chickadees. Each one sits on a designated twig, impatiently awaiting its turn to swoop in and snatch a sunflower seed. Back on the branch, she will grasp the seed with dexterous toes while banging it with her stout little beak. A month from now, she will put that beak to good use chiseling out a nesting cavity in one of the nearby birch trees.

Occasionally the calm is broken by the screech of a red-shouldered hawk. The chickadees scatter and regroup beneath the snow-covered skirts of a neighboring hemlock. They have been fooled again, this time by a flock of blue jays. The crafty birds can mimic the calls of predators and seem to enjoy scaring the hapless little birds. Afterwards, they hop and bow, emitting curious little mews and burbles, as if they are congratulating each other on their extreme cleverness. Finally, the blue jays mob the feeder like unruly children on a Halloween night. They grab twenty seeds at a time and scatter them in the snow. Perhaps they are dispersing them for future use, perhaps they are just pigs. Eventually, they whiffle through the turgid woods, burbling as they go . . .

I'm sure I could learn a lot if I stay to watch this feeder. But I am struck with this urge to find some green; not the black green of conifers, but the verdant green-green of new life. Besides it is close to Saint Patrick's Day, surely excuse enough to mount an expedition in search of green.

Hopefully I will discover some small sign of growth, some inkling of spring, some reason for hope. But it is largely no go. Coal Hill is a world of monochromes: gray skies, black bark, white snow. Even the needles of the conifers look more black than green today. I pause to inspect fresh pieces of a spruce cone scattered in the snow. The perpetrator scolds me from a nearby branch. It is another red squirrel evidently on his own quest for food.

But suddenly I notice what I've been looking for: delicate, luminous, translucent shades of green. They glimmer through the leafy lobes of a lichen growing on the bark of the red squirrel's spruce tree. In summer, these lichen are crusty, dry, and easy to ignore. Today they are moist, swollen, and suffused with these delicate hues.

Lichen are fascinating organisms, but they harbor a twisted past. We can think of them as plants that didn't quite make it. Instead of evolving into a single plant with chloroplasts, lichen are a symbiotic partnership of fungi and algae. Under a hand lens, I can see the fungus holding onto the alga

A world of monochromes. Courtesy *The Courier* newspaper.

Lichen,
Tuckermanopsis *sp.*

with hundreds of translucent gray threads. The photosynthetic alga provides the fungus with food, while the fungus provides the alga with support, structure, and a moist replica of the ancestral ocean from which it evolved.

In the tropics, corals do a similar thing. Zooantellae algae provide the corals with photosynthesis, while coral polyps provide the structure to hold and protect the algae from predators. But how do these symbiotic partnerships come about and why are they so prevalent?

If we go to a rocky coast, we can see some clues. At the very top of rocky shores, you have what is called the splash zone, which is dominated by Xantheria lichen. Just below the splash zone is the black zone, which is covered with a thin film of algae that exude a coating of dark slime for protection from the sun. But be careful, the black zone is a treacherous area slippery to walk on when the slime is moist. It may also be treacherous for lichen; in places they are overgrown with the algal slime.

Why would anything want to live in either of these zones? Predators. There are none to speak of in this no-man's-land that is too dry for most marine organisms and too salty for most land creatures. Occasionally, you might see a periwinkle cruising into the black zone to scrape away a few algae, but all in all, it is a predator-free environment.

So it would certainly make sense for an alga to move out of the predator-infested waters of the coast and up onto the predator-free rocks. It would make even more sense for it to team up with the nearby fungi. The fungus

Symbiosis on a worldwide scale; lichen on lava, Iceland.

could supply the alga with water and protection from the sun, and the alga could supply the fungus with the ability to grow its own food. Together, the two could move off the coast and spread into the continent. Their partnership would have allowed them to be among the very first organisms to move out of the ocean and onto the land.

Today, lichen continue to thrive in those places too harsh for true plants. In doing so, they provide the world with a valuable service. They can repair the biosphere after geological change. When an ice age scours away the topsoil or a volcano covers up an area, lichen are among the first organisms to arrive by airborne spores. In as short a time as fifteen years, they can break down boulders and build up soil, preparing the area for true plants. Did the early lichen do the same thing on land, inadvertently preparing the terrestrial world for life? When you think about it, such a service can even be considered symbiosis on a worldwide scale.

Indeed, photosynthetic algae are one of evolution's most valuable inventions; so precious that numerous organisms have gone to great lengths to forge symbiotic partnerships with them. The benefits are obvious: the partnerships allow the new organisms to grow their own food. They are so effective that they arose at least three times; once on land with lichens, once in the sea with corals, and once in cells with plants. It is probable it happened many times more.

It is difficult to imagine our planet without such partnerships. It is doubtful we would even have had life at all, except for a few isolated communities of chemosynthetic creatures clustered deep in the ocean around underwater volcanoes. It is certain we would have a lot less green.

One of the most difficult steps that scientists have yet to explain is how life evolved from complex molecules into replicating cells like those found in blue-green algae. Indeed, some scientists have simply thrown up their hands and said, "Cellular life is just too damn complicated. There simply wasn't enough time for it to have originated on earth. It must have come from space, . . . or God."

Scientists can't speak for the latter. But if life did come from space, I can't think of a better way to get things off to a good start than to inoculate a planet with blue-green algae. Perhaps our algae did arrive as cysts in asteroids such as those from Mars that were recently found in Antarctica. However they came about, one thing is clear: Once you have blue-green algae, they are simply too important to leave alone. Soon, dozens of organisms will want to incorporate them into symbiotic partnerships, and millions of species will evolve from the useful liaisons.

Some scientists have even suggested that we should ship blue-green algae to appropriate planets to prepare them for life, oxygen, and human occupation. The idea is not without chutzpah. Of course, all we might do is snuff out the planets' existing early life by introducing noxious oxygen pollution.

Lichen are currently in the midst of a paleontological argument. Scientists have long been mystified by the Ediacaran fossils, believed to have been from soft-bodied marine animals that appeared before the Cambrian explosion. Scientists, including Stephen Jay Gould, have argued that they were just early animal forms that didn't make it. Recently however, another paleontologist has suggested that the fossils might really be from lichen. This would explain why they were so well preserved and how they existed before much food was available.

If the argument is true, it would mean that lichen evolved their partner-

The skier's nemesis, Tuckerman's Ravine, was named for a Harvard lichenologist.

ships with algae underwater, and even earlier than we thought. They may have been the first organisms to invade the land after the Vergannes Ice Age that almost obliterated life before it got started. If so, they could have been an important first step in preparing the earth for true plants.

Numerous people have had a go at unravelling such secrets about lichen. The lichen on my spruce tree, *Tuckermanopsis*, is named for the same scientist who gave his name to Tuckerman's Ravine, the skier's nemesis on nearby Mount Washington. Tuckerman was a prolific taxonomist. Not only did he have a lichen genus and a ravine named for him, but you can still see the Tuckerman collection of lichen at Harvard University.

Another pioneering lichenologist received much less recognition for her work. So little, in fact, that she was forced to write children's books to make a living. But perhaps the world has not been so poorly served by the loss of yet another fungologist as it was well served by Jeremy Frog and Peter Rabbit, for Beatrix Potter was that most delightful of all creatures of symbiosis, a partnership of scientist and artist.

Sunday on Sugar Hill. Courtesy *The Courier* newspaper.

Chapter 10

Sugar Hill

MARCH 18, 1999

I hear a new sound in the forest this morning. It is the welcome, restless sound of spring in the North Woods. A horse paws the snow and jingles the bells of his heavy harness. A farmer quietly moves from tree to tree tapping in spouts, checking plastic tubing, stoking his sugar shack fire. Soon the sweet smell of maple syrup will waft through these trees.

It is Sunday on Sugar Hill. The days have been getting progressively longer since December 21. Now the temperatures consistently rise above freezing by day, but dip back into the twenties at night; perfect conditions for tapping into the sweet sap of sugar maples.

However, the farmer is tapping into something far greater than maple trees. He is tapping into the energy of the universe. All life must use energy to battle the forces of entropy. Without it, matter would simply run down into chaos. It's an uphill battle, this insistence of life to become more and more organized. The way plants do it is by tapping into the sun's energy.

It is an ability based on an ancient partnership. Two and one-half billion years ago, the earth was still a planet of small proto-continents, but the surrounding seas were teeming with single-celled organisms. Eventually, some of the photosynthetic algae started to parasitize bacteria cells. Initially, the situation was fatal. Oxygen from the alga caused the tissue of the bacteria to combust. Over time, however, the bacteria learned how to use the oxygen

Testing viscosity. Courtesy *The Courier* newspaper.

for aerobic metabolism. It was a remarkable partnership—so powerful that it polluted the atmosphere with oxygen, almost snuffing out life before it had hardly begun.

Life recovered, of course, and the symbiotic creatures became the chloroplasts in plants, the organelles that make photosynthesis possible but still contain their own DNA and RNA. It is these chloroplasts that allow trees to use the sun's energy to pull carbon out of carbon dioxide and combine it with hydrogen and oxygen to make sugars.

Trees do this better than most plants and maples do it better than any. In only three short months last summer, these maple trees converted enough energy into sugars to fuel the blossoming of flowers, buds, and leaves for the coming season. They did it so exuberantly that humans can tap into their reserves without harming the trees or affecting their following year's growth.

But it is an all-or-nothing strategy. The maples require perfect conditions: access to lots of sunlight, nutrient-rich soil, and excellent drainage. They place all their bets on being able to produce all the sugars they need during their short growing season. They do this by growing large, highly efficient but delicate leaves that cannot withstand winter conditions. Evidently the strategy serves them well. Maple sugar is a truly renewable resource.

Below the maples are trees that have adopted a more conservative strategy. They are the conifers. These evergreen trees have hardy needles that can survive winter so they don't have to go through the costly process of regrowing leaves every spring. This gives the conifers a headstart on the growing season. They have already been photosynthesizing for several weeks, but the maples won't start until they leaf out in May.

Needles, however, have fewer stomata, the tiny leaf pores that trees use to pull carbon dioxide and water out of the air. This makes conifers ten times less efficient than hardwood trees. But conifers make up for their lower efficiency by requiring less perfect conditions, so they can live in colder, wetter, shadier conditions.

It is a silent but relentless competition. Conifers have the advantage when conditions are harsh, as they were after the retreat of the glaciers. Fast-growing hardwoods have the advantage when conditions improve. Maples are quick to colonize new patches of sunlight like these at the edge of the farmer's field. However, their offspring will not be able to survive in the shade of the young conifers now biding their time in the understory. The area will revert to the conifers. Such are the hard ways of natural succession.

But now the early morning calm is broken by a fusilade of wings. A ruffed grouse has burst from beneath the snow. Last night, he flew into the snowbank to prepare his evening burrow. The snow helped insulate him from the colder surface temperatures and hide him from predators.

I peer into his burrow, and spot two dark green, delicate shoots poking through the snow. They are lycopods, primitive plants that have taken the conifer strategy to the extreme. They look so much like conifers that they are often called running pine or ground cedar. They are actually two species of club mosses that rise only four to six inches above the forest floor.

Maple sugar buckets.

But it was not always so. The ancestors of these species would have towered over any tree in this forest. They were Lepidodendrons, giant trees that rose 150 feet over the swampy, fertile, amphibian-rich landscape of the Carboniferous period.

The early lycopods were locked in an arms race as relentless as the one facing today's hardwoods and conifers. But their competition involved several more players. The trees were growing taller to avoid being eaten, the herbivorous dinosaurs were growing bigger to avoid being consumed by carnivores, and the carnivores were growing bigger to bring down the herbivores. By the end of the Mesozoic, huge, fast-paced *Tyrannosauros rexes* were chasing down lumbering *Brontosaurii*. Much vegetation was trampled in the process.

But hidden below all this Fantasia-like carnage were a few mutant lycopods that had adopted a different strategy than their gigantic relatives. These dwarf lycopods lived low to the ground, using only the meager amount of sunlight that filtered through the towering canopy of their overgrown cousins.

During the time of the dinosaurs, the strategy allowed the dwarf lycopods to avoid the unwanted attention of herbivores. Later, it allowed them to avoid extinction. At the close of the Mesozoic era, a giant asteroid plunged into the Yucatan, throwing up thick clouds of dust and smoke from burning trees. The clouds curtailed photosynthesis, so that all the dinosaurs and ninety percent of the creatures in the ocean died out.

But what about the little lycopods? They survived, and continue to flourish on the small amount of energy that trickles down to the understory. Like the conifers, they keep their needles all year long. Unlike the conifers, they can happily survive winter beneath the protecting mantle of the snowpack. For them, less is surely better; they can even photosynthesis on the weak blue light that filters through the snowpack.

Lycopods poking through the snow.

Lepidodendron forests in the Carboniferous Age.

The lycopods' growing season is several weeks longer than conifers, plus they don't have to expend any energy to compete with trees to grow above the canopy. The snowpack protects them from the dangers of ice storms and heavy snow. They have even lost the ability to make wood. Who needs it when you can live quietly under the snow? Theirs has been a slow, graceful evolution to modesty from their former noble heritage. It is probably a good thing too. The coal the farmer is using to bank his stove is made from the carbonized remains of the once-noble Lepidodendrons. Does the evolution of the lycopods hold a lesson for people and countries in similar circumstances?

The farmer drizzles a splash of syrup into the snow. I try it. I'm truly tasting the energy of the universe. Last summer, energy for this sugar traveled ninety-three million miles from a nuclear reaction in our sun. The carbon in this sugar was made from a former star that exploded out of a supernova. The hydrogen was made seconds after the Big Bang. The water has been knocking about this planet for at least four billion years.

These mountains are covered with thousands of species of trees. Each has devised ingenious ways to harvest the sunlight, water, and carbon dioxide that sweep over our planet. In just this one taste of maple syrup, I might be ingesting hydrogen from a whale's breath in the Atlantic, oxygen from a toucan in Brazil, carbon from the tailpipe of a taxi cab in downtown New York.

We're all in this soup together, part of a great cosmic experiment; trad-

Maple tree with old bore holes. Courtesy *The Courier* newspaper.

ing, using, recycling chemicals at a dizzying clip. This is the stuff of the universe that I'm drinking.

Tonight when I build my fire, I will once again bask in the glow of the universe. This time the energy will be from wood, carbon, hydrogen, and oxygen reassembled into the form of lignin and cellulose. I find it a warm, connected, humbling thought.

PART III

Spring

> *R.W.E. (Ralph Waldo Emerson) tells me he does not like Haynes as well as I do. I tell him that he makes better manure than most men.*
>
> —Henry David Thoreau, *Journals*, May 4, 1852

Squaw Moccasin. A tribal leader of the Wampanoags quietly corrected me once for calling this plant a lady slipper.

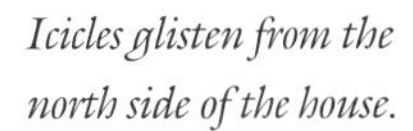

Icicles glisten from the north side of the house.

Chapter 11

The First Day of Spring

> *No doubt he liked to scorch off*
> *morning fog by simply staring through it*
> *long enough so that what he saw grew visible.*
> —William Mathews, "On the Porch at The Frost Place"

MARCH 1999

It is a week before the vernal equinox. The sun continues its slow progression north but the weather ricochets wildly between the conditions of winter and spring. Eighteen inches of new snow have fallen on the mountains, but the temperature is already above freezing and rising fast. Icicles glisten from the north side of the house and a curve of snow hangs off the barn roof like a wave poised to collapse.

I move a kitchen chair onto the porch and sit down to enjoy the morning. My daughter is playing out of sight at the bottom of the hill. A German shepherd gambols onto the meadow, throwing his head back in wild abandon. I notice him out of the corner of my eye, but pay little attention. He belongs to our neighbors. They are probably behind him, coming up the hill for an early morning walk.

Suddenly it registers. That's not a German shepherd. It's a bear cub. His mother must be close at hand. I rush for my camera and yell for my daugh-

Snow curves off the tin roof like a wave about to collapse.

ter. It's too late. The cub has disappeared. My daughter runs up the hill. No photograph, but a safe child. I'll settle for the exchange.

I clamber down the hill to a mountain stream that tumbles through our meadow. Filaments of green algae sway back and forth in its gentle current.

Large gray flies lumber slowly over the new-fallen snow. Some are drying their wings on the boulders beside the stream. These are the adult forms of stoneflies whose nymphs have just hauled themselves out of the stream to hatch.

The adults will quickly mate and lay their eggs back in the stream. This late-winter mating allows the stoneflies to beat the hordes of avian predators that will arrive in a few weeks. It will also give the larvae time to grow and terrorize the smaller creatures of the stream. I hope they will feast heartily on blackflies.

A tiny subnivean tunnel disappears into a snowbank near the stream. It probably belongs to the white-footed mouse whose squeak so startled my daughter when she stepped on its burrow last summer. Beneath the snowpack, trout lilies are already sprouting. They are able to photosynthesis in the blue light that filters weakly through the snow. Their delicate flowers will soon grace the forest floor, for they must bloom before the leaves of deciduous trees block the sunlight.

Even these naked goldenrods harbor life. Wasp larvae will soon emerge from galls on the goldenrods' spindly stalks. The larvae must be using a biological clock to keep track of time over 365 days. Sequestered in the galls, they receive no outside clues such as increasing light to trigger their emer-

Wild lilies. Courtesy *The Courier* newspaper.

gence. Humans lose track of time within twenty-four hours, without outside cues to recalibrate our biological clocks.

Each species of plant and animal uses different cues to trigger its springtime activities. Some use the lengthening amount of daylight, others the rising temperatures. The result is a seemingly well-orchestrated unfolding of life and behavior. But few dare rely on the vagaries of springtime weather. One exception is the bear.

I have found the cub's tracks, and follow them back into the woods. They are joined by the tracks of the mother bear and another cub. Judging from the tracks, the mother bear gave the errant cub a good cuff before herding it back into their den beneath an overturned spruce tree. I've passed the den a dozen times, never guessing it contained the sleeping family.

Bear scat looking vaguely like a dollop of raspberry jam.

The mother bear has a world class case of constipation. Charles H. Willey.

The mother bear is undoubtedly the same bear we saw last summer. She consorted briefly with a young male who had wandered into her feeding territory to gorge on raspberries. We had seen their scat. It looked like someone had emptied a remarkably large jar of raspberry jam into the grass. Later, she feasted on apples and rested in the sun. I like to think the pair lay together after mating, but the young male was probably long gone.

Last October, the female excavated this den in the cavity beneath the overturned spruce. She had feasted on nuts and berries, fasted to clear out her digestive tract, then ingested tufts of her own hair to block her rectum for winter.

Finally, she dozed off. Her metabolism slowed to half its usual rate, but unlike true hibernators, she had maintained her body temperature while sleeping. This allowed her to wake and even walk outside on warm days.

In February, the female bear was barely awoken by the birth of her two cubs. It was an enviable birth: The cubs only weighed eight ounces, little more than a stick of butter.

Now the cubs are plump and healthy. They will rely on their mother's milk for well into the summer. Not so the unfortunate mother. She is hungry, thirsty, and has a world-class case of constipation. I decide it is best not to make her acquaintance under such circumstances.

I return to the house to tell of my adventures. They elicit little interest until I come to the part about the bear's birth. My wife seems particularly enamored of the idea of sleeping through the birth and first three months of her offspring's life. Remembering our own sleep-deprived existence during similar times, I'm hard-pressed to disagree.

But now the weather has shifted, and thick gray clouds shroud the valley. We decide to explore by car. People are using these last few days of winter to finish up seasonal chores. On the Easton Road, our car is enveloped in sweet smelling clouds of steam from Charley Stewart's maple sugaring operation. We pass Robert Frost's old farm. I imagine him standing on the porch scorching off the morning fog with withering blasts of outrage. Beyond the farm, a few immigrant flatlanders have tried their hand at tapping maples.

The stone chimney of the Franconia Iron Works glowers from the edge of the Gale River. The river was dammed to provide power for the two giant bellows that made the charcoal burn hot enough to turn ore into iron. Further up the river, another dam powered a sawmill that produced lumber for the town and surrounding communities. With just the power of horses, men, and the river, these early industries denuded the area of trees. The chimney always reminds me of the time when logging was a low-tech but still devastating operation.

Up the road is an example of a modern logging operation. The crew is hurrying to clean up a site while the ground is still frozen. The area has the piney smells of a lumber yard but the look of a construction area. A huge crane grasps half a dozen trees from the top of a fifty-foot pile. It hands them

On such days I imagine Robert Frost standing on his porch scorching off the morning fog with withering blasts of moral outrage.

The stone chimney of the Franconia Iron Works glowers from the edge of the Gale River.

to a mechanical grappler, which feeds them into the waiting maw of a chipper. In an instant, the thirty-foot trees are reduced to shreds and shot into the back of an eighteen wheeler. The chips will be trucked to a power station to be burned to produce electricity. Formerly, this slash would have been left in the woods as an eyesore and a potential source of fuel for forest fires.

On the other side of the loading area, piles of logs are neatly sorted by species. The larger pines will be lifted onto flatbed trucks and shipped to sawmills for lumber. The smaller logs will be trucked to papermills to make pulp for paper. The birch will be used for veneer, and the other hardwoods will be cut for firewood.

I'm impressed with the operation. The men are proud of their neat work and costly machinery. They are particularly appreciative of the safety of

A huge crane gathers up half a dozen trees. They will be chipped and used as fuel to produce electricity for the New England power grid.

Turkeys are wiser than they look. This flock led me on a mile-long chase keeping just out of camera range.

these modern methods. Most of them can remember relatives who died in horrible logging accidents. Today, loggers hardly have to come in contact with logs and saws. They sit inside the sturdy, metal, safety cages of machines that look like mechanical dinosaurs. A feller-buncher can grapple a seventy-five-foot tree, saw off its base, lift it overhead, and hand it to a skidder that drags it to a distant landing area for delimbing.

What's left behind is a diversified forest "weeded" of its mature trees and ready to grow more. I find such a forest far more appealing than monoculture tree plantations that must be clearcut, herbicized, and planted for the next rotation.

By late afternoon, the weather has turned again. Clouds have cleared in the west and the setting sun casts a golden glow on mountains still backed by thick, gray clouds. As we pull into the driveway, I look down at the meadow to spot a flock of turkeys scratching in a small clearing beneath a pine tree. They are descendants of twenty-four wild turkeys that were reintroduced into New England in 1975.

The turkeys rush into the woods, but double back onto the road. I follow them until I'm blocked by snow. I jump out of the car, but keep it idling; to shut it off might startle the birds. The four females trot up the hill, jostling and gobbling at each other like shoppers shoving to be first in line for a Filene's Basement sale.

But the turkeys are wiser than they look. The male is larger, with prominent wattles. He positions himself between me and the flock and hurries them on to ensure that they always stay just out of camera range. We trot up the hill this way for over a mile. At the top, the turkeys scuttle into the underbrush and disappear.

Moose use their lower teeth to browse bark off of trees.

The first day of spring, Mount Lafayette.

I am left to investigate another winter kill. Blood marks the snow, but I also see curious speckles. I pick one up to investigate. They are the scales of a six-inch fish. I'm hundreds of yards from the nearest pond or stream, so it is probably not a muskrat. The scales are probably the remains of a meal left by an otter that has taken up winter residence in our neighbor's pond.

Nearby are deep scratches on a maple. It looks like they could be made by a bear, but on closer examination I see that they are made by the lower teeth of a moose. She used her teeth to strip the bark off the young tree. It's remarkable to think that a hundred years ago I wouldn't have seen any moose, deer, turkeys, or bears in New Hampshire. Today I have seen signs of all of them.

But now it is time to head home. I look to my left, and see the last golden rays of the sunset reflecting off the snow-capped peak of Lafayette. The sky behind is still a deep, dark gray. The light changes from pink to salmon to blue. The shadows climb the flanks of Lafayette until it is dark.

Soon the snow cross will appear. It appears on the flanks of Lafayette as the snow melts on most of the mountainside, but remains in two ravines that intersect. Around here, the snow cross is considered the first true harbinger of spring.

I walk down Cole Hill to find that my car has been idling for several hours. But it has been worth it. Even though it is still a week before the equinox, in my book this will go down as the first day of spring.

The vernal equinox; sunrise on pine trees.

Chapter 12

The Vernal Equinox

A Springtime Digression

MARCH 20, 1999

Today is a remarkable day. From the North Pole to the South Pole, the entire world is briefly united in experiencing twelve hours of daylight and twelve hours of darkness. It is the vernal equinox, the first day of spring. We usually celebrate the season with bacchanalia—often war. This year it is the tragedy in Kosovo. Does nature celebrate the season in similar ways?

Yesterday, I heard a robin on the top of Coal Hill. A month before, the increasing length of daylight had signaled her hypothalamic-pituitary system to secrete hormones that increased her energy and alertness. Her flock had become agitated and restless, gripped with a neuronally induced mass hysteria. At the slightest provocation, her flock would jump into the sky, circle about frenetically, then alight once more. When the agitation finally reached its peak, the entire flock rose up to start its arduous migration north.

Last week, rising temperatures awoke the mother bear and her cubs. But what about humans? Surely we are part of nature. Do we use the strategy of the bear or the strategy of the robin to trigger our springtime activities? Are we aroused by rising temperatures or lengthening daylight?

I find several clues. First, let's look at our springtime celebrations: March

Madness, spring riots, May Day. They have secular, religious, and political overtones, but ultimately they ritualize biological urges.

The festivities become more extreme the further north you go. Inuit tribesmen traditionally go berserk in the spring. The word "berserk" itself comes from the name of Finnish warriors who would embark on rampages of rape, violence, and drunkenness at the end of their depressingly long winters. Could it be that what we call manic depression or bipolar disease arose out of humankind's move to the North?

Let's look at manic depression's more benign cousin, seasonal affective disorder, or SAD. In New Hampshire, 34 percent of the population suffers from SAD, compared to less than 5 percent in Florida.

When the public thinks of SAD, it thinks of winter depression, which is to be expected; that's what people with SAD complain of. However, the main feature that researchers use to diagnose SAD is whether a person has hypomania, the summertime elation that makes SAD sufferers more aggressive, promiscuous, and sometimes downright obnoxious in summer. As someone who has SAD, here I speak with a certain amount of authority.

Of course, hypomania presents a double-edged sword. We SAD sufferers love our summertime highs, our flights of fancy, our increased productivity. Most scientists now place seasonal affective disorder on the lower end of a continuum with manic depression. The psychologist Kay Jamieson has shown that an unusually high percentage of both disorders appears in creative people, and that the afflictions somehow add to their creativity.

The vernal equinox, sunrise on birch trees.

Hibernation, the strategy of the bear.
Mass. Wildlife/Bill Byrne.

Artists seem to be particularly sensitive to the changing light of the seasons. Think how many artists have tried to capture the different light of the seasons. Think of how many musicians have tried to capture the seasons' moods.

I have a personal writer's trick that capitalizes on my seasonal affective disorder. When I am stuck on a particular piece of writing, I try to think of how I would feel on a specific day in a specific season. Once I relive that mood, my brain seems to change gears and the words start to flow.

Dr. Jamieson has also shown that creative people with SAD produce up to 80 percent of their work when the light is changing the fastest in the spring and fall. I find this to be true, but also feel that my writing is a form of self-medication; it allows me to forget that I'm depressed in the winter and keeps me out of trouble in the summer.

Of course, this system can get sticky. Manic depressive businessmen have been known to buy too much or sell too much in spring, politicians can lead us all too willingly into springtime wars, and artists can become too downright talkative, hypersexual, and obnoxious for their own good.

So, does SAD represent a form of hibernation evolved recently to cope with winter—the traditional view based on the model of the bear? Or, does SAD represent a glitch in an ancient system evolved to get the maximum benefit out of living in the north—the model based on bird migration?

To find out, let's take a brief detour into my brain. When I awoke this morning, the spring light triggered an electrical impulse that traveled down my optic nerve to my brain's pineal gland and hypothalamus. The hypothalamus, in tandem with the pituitary gland, initiated electrical signals and hormones that cascaded through my sympathetic nervous system. The sympathetic system triggers such things as action, aggression, and mating. Its most famous behavioral repertoire is the well-known fight or flight response in humans—in birds, it's migration.

The pineal is the gland that makes and breaks down serotonin and melatonin. At night, darkness triggers the pineal to break down serotonin and make melatonin, so we sleep. In the morning, the process reverses: light triggers the pineal to break down melatonin and reassemble it again as serotonin. We wake up and are alert and active. Serotonin is also the neurotransmitter most associated with depression and elation.

This system is the reason that light therapy works faster and is cleaner than antidepressants for people with SAD. It is faster because it initiates the system at the top of the cascade; no time is wasted waiting around for serotonin to build up in the synapses between nerves. It is cleaner because it targets this specific system rather than taking a more generalized, central nervous system approach that would produce unwanted side effects.

This simple model shows how such a system can provide people with more serotonin in the spring and summer and less in the fall and winter. It also shows how a minor glitch in the system could lead to seasonal affective disorder, and a major glitch could lead to full-blown manic depression.

I hope the reader has excused this brief voyage into my personal neuropathology. I happen to be the person I am most familiar with who has sea-

Blue jay. Mass. Wildlife/Bill Byrne.

The human brain has evolved a system to reap the benefits of moving away from the never-ending equinox of the equator.

sonal affective disorder. I think it casts light on life in the North for both humans and other animals.

But how did this system come about? Is the agitation and restlessness that birds feel prior to their migration similar to what seasonally affected humans feel prior to the onset of hypomania?

The benefits of birds' migrations are easy to calculate. In New Hampshire, a pair of robins can raise their chicks from hatchlings to fledglings in ten days; in Washington, D.C., it takes them about thirteen days. That is a crucial margin. It decreases by a third the amount of time that flightless chicks are most vulnerable to predation. It increases by a third the chances of raising more chicks.

Those extra chicks are what makes migration worth the dangers and effort. Otherwise, it would make more evolutionary sense for birds to stay in the tropics where the climate is mild and food plentiful all year long.

But how did bird migration evolve? We can assume that birds used to live in the North and earlier ice ages pushed them toward the equator. As the glaciers retreated, it made evolutionary sense for them to migrate back to the North for the greater amount of summer daylight and south for the mild winter climate. Over time, they developed what is called a stable evolutionary strategy.

Our early hominid ancestors faced similar pressures when they radiated away from the equator more than a million years ago. It made evolutionary sense for them to evolve a system that allowed them to reap the benefits of being more energetic and extroverted in summer and the benefits of preserving energy and having the introversion and concentration to work on detailed chores in winter.

Humans who became more active, persuasive, and promiscuous in the summer would increase their chances of siring more offspring. So manic depression and SAD can be thought of as glitches in a natural system that evolved not since the ice ages, but millions of years before. They remain in the human gene pool because mild forms of manic depression and SAD can still confer benefits to ourselves and our offspring.

Our early ancestors had evolved a system that allowed them to change their personalities to match the physical and social environment. In essence, they evolved a system to reap the benefits of moving away from the never-ending equinox of the equator.

Patches of ice still linger in the forest.
Courtesy *The Courier* newspaper.

Chapter 13

The Vernal Pond

APRIL 17, 1999

It is a cold, wet night in early April. Raindrops patter quietly on the forest floor. Patches of ice still linger in the shallow swales used by moose in their sleeping yards. Beneath a pile of leaf litter, a wood frog, *Rana sylvatica*, prepares to reenact a ritual that sends shivers down my spine, for it was on nights like these that our mutual ancestors first walked on land.

Last autumn, the wood frog dug into this leaf litter to hibernate. Her temperature had dropped in synchrony with the surrounding earth. But, at just about freezing, a strange thing happened. Latent heat released by the nucleation of ice crystals in her body fluids raised her temperature and kicked her heart into overdrive. It pumped furiously for several hours, saturating her tissues with glucose. She became severely diabetic, but protected by the antifreeze properties of the potent sugar.

Twenty hours later, her heart stopped, she ceased breathing, and switched to anaerobic metabolism. She was teetering on the very edge of life, for a drop of only a few degrees below freezing would overwhelm her glucose. She could freeze solid and die.

Snow saved her, however. It fell heavily during autumn, and its insulation kept her hiberniculum just a few critical degrees below freezing. Winters without snowcover can kill millions of hibernating amphibians. This is another thing to ponder in our age of global change.

But several hours ago, the temperature rose above freezing and the wood frog's heart started beating once more. It flushed the glucose from her bloodstream, then resumed its normal rythmn. An hour after thawing, she was ready to mate. She will to continue pass back and forth from frozen to mating for many more nights in this early spring. It is a remarkable adaptation that allows her to mate and lay eggs before most of her predators.

Now the frog digs her way out of the ground to join a slithery migration of amphibians hopping and crawling toward the cold, dark waters of a vernal pond. The route reverses the migration of our mutual ancestors, the tetrapod amphibians. Three hundred and sixty-five years ago, the tetrapod

A frozen wood frog. In the morning they thaw, mate, then freeze again for the following night.

amphibians crawled out of the water to escape the sharklike predators that terrorized the rivers and estuaries of the Devonian.

The ancient amphibians had already evolved limbs to help them hide in underwater vegetation. But they found their limbs especially useful once they hauled out on land. It was a land of fern and lycopod forests, happily devoid of predators. It was an amphibian Eden filled with food—if your tastes ran toward dragonflies and cockroaches the size of small rats.

Like their ancestors, today's amphibians—frogs, newts, and salamanders—still must return to water to lay their eggs. They prefer vernal ponds that fill only during spring rains, because the temporary ponds harbor fewer predators.

As the frog enters the pond, she must swim through a gauntlet of lascivious male amphibians. The spotted salamander males have already placed packets of sperm on the underwater vegetation. The female salamanders will pick up the packets and tuck them into their cloacae for insemination.

The female frog is not so lucky. She must make her way through a deafening stagline of peeping, bellowing, quacking males. The spring peepers make the most noise, but to the female, wood frogs have the sweeter calls. The wood frogs sound like a flock of tiny feeding ducks, the diminutive tree frogs like broken guitar strings.

Now a male courts the female with a tentative foreclaw. She kicks him away. Another male approaches. She lets this suitor mount. Together the two struggle toward a patch of pondweed. She clasps the weed and extrudes her eggs while the male fertilizes them one by one.

Well hidden among the rocks, a red newt.

The pond is going through its own metamorphosis. Courtesy *The Courier* newspaper.

Hundreds of other frogs are doing the same thing. By midnight most have left the pond and returned to the forest. But they will repeat the process many more nights as the temperature fluctuates above and below freezing.

But the frog has left her eggs in the pond oblivious to the presence of a red-spotted newt. It has done the frogs one better. It started as an egg, hatched as a gill-breathing larva, metamorphosed into a land-living eft, then metamorphosed again into a swimming three-year-old adult—an adult now intent on raiding the eggs of other amphibians while preparing to lay her own.

The newt will be able to wrestle a few frog eggs out of their gelatinous coating, but most will survive to become full-fledged frogs by mid-summer. Not so the tadpoles of the bullfrog, which can take up to three years to metamorphose.

The pond, meanwhile, has been undergoing its own metamorphosis. All winter long it lay sequestered under a lid of solid ice. The ice had protected the pond's creatures from sub-freezing temperatures but had cut them off from oxygen and light. It had been a long and dangerous winter.

With reduced light, phytoplankton had curtailed photosynthesis, but animals had continued to respire and decomposers had continued to break down organic matter. Oxygen had declined to dangerous levels, while nutrients had accumulated on the pond bottom. The last few weeks of winter had been a game of endurance. A few creatures had already died.

But, ever so slowly, the spring sun started melting the slushy ice surface, waking the pond from its winter torpor. As the ice thawed, its meltwater plunged toward the bottom, for water is densest just a few degrees above freezing. Now even the slightest breeze can churn up the pond, saturating it with oxygen and stirring up clouds of nutrients.

This is the spring overturn. It makes spring underwater as dramatic as that on land. One day the water is crystal clear. A few days later, it will be filled with greenish, bluish clouds of phytoplankton. A few days later, the water will be full of zooplankton, tiny drifting animals that are feeding on the rapidly reproducing phytoplankton.

I can see a miasma of zooplankton flickering in the beam of my flashlight. I see amphipods, copepods, ostracods, even decapods. The spring overturn has transformed this pond into a vast nursery. Each day, a new species hatches to feed on organisms that reached their peak of abundance a few days before. It is a well-orchestrated unfolding, finely attuned to the rising temperature of the pond. Underwater, temperature, not increasing daylight, is the most reliable clock by which to time your springtime reproduction.

Not so on land. The morning sun now shines on eight robins that used the increasing daylength to trigger their migration north. They pause to listen for worms in the newly sun-softened earth. I watch the same flock of turkeys feeding on the brown stalks of sensitive ferns, while a ruffed grouse picks its way through the understory. The flock of turkeys will soon disperse and the females will start to nest.

But the ferns will reproduce in a more elaborate way. They were one of the first plants to invent sex, and continue to reproduce both sexually and asexually. The brown stalks contain spores, asexual clones of their parents. Soon, millions of spores will blow out of the dried stalks and disperse over the land. Some will land on moist soil and germinate as a single-celled prothallus. The prothallus will add new cells but will continue to look like nothing more than a tiny green growth.

However, the prothallus is actually a small plant that will produce sperm and eggs. The sperm are covered with hairy cilia and shaped like corkscrews. They swim through moisture on the prothallus to fertilize its eggs. The eggs

Fiddleheads, the asexually reproducing form of adult ferns. Courtesy *The Courier* newspaper.

Jack in the pulpit.

will then grow into fiddleheads, and then into the delicate green fronds of the asexually reproducing adult ferns.

This alternation of generations allows us to see how sexual reproduction evolved. But why did it evolve? During times of stability, it makes more sense to reproduce asexual clones already adapted to the unchanging environment. During times of change, it makes more sense to reproduce sexually, and increase your chances of producing offspring better-suited to a new environment. In essence, sex speeds up evolution—to say nothing of making life a little more interesting and a lot more complicated.

We can see just how complicated sex can get by looking at a patch of snow at the edge of the vernal pond. A bulbous purple-and-green mottled spath juts suggestively from the forest floor. It is surrounded by the large green leaves of a skunk cabbage. The whole scene has the fetid smell of putrefying flesh. It is all part of an elaborate sexual deception. The skunk cabbage has produced this odor to attract tiny gnatlike flies to its pollen. It has completed the hoax by generating heat to melt the snow so it looks like a small corpse is rotting near its leaves.

Plants evolved the ability to deceive animals into pollinating for them during the Cretaceous era 144 million years ago. It was a neat trick, a vast improvement over methods used by the ferns and lycopods during the swampy years of the Carboniferous, a hundred million years before.

Fortunately, skunk cabbages are not as proficient at producing a stink as the Amorphophallus of Sumatra. People have been known to pass out from simply walking under their fetid blooms, if not faint from embarrassment at the sight of their suggestive flowers.

Not far away, I come across a large dried-out scat. It is full of white hairs, fragments of large bones, and tough horny material. It seems catlike, but it is obviously too large to be from a bobcat. Lynx have been extirpated from this area and reports of mountain lions are said to be unreliable . . .

Coyote scat.

I slip the scat into my pocket and hasten down the mountain. In some soft mud, I spot four tracks of a running animal with long, large claws. They seem fully capable of killing a deer. The thought of a mountain lion niggles at my limbic system. Mountain lion have been known to hunt humans and attack horses. Could one of them be silently tracking me? The forest seems suddenly more foreboding.

Finally I reach home, and call the U.S. Forest Service. We go through the characteristics of the scat. I'm careful not to mention my suspicions. The specimen is four inches long and bulbous. That rules out an owl or bobcat. The scat has fragments of large bones and horny material that looks like it could have been gnawed off a hoof. The hair is short, so it is probably from a deer, not a snowshoe rabbit. That rules out a lynx. That leaves nothing but a mountain lion.

I can hear the excitement on the other end of the phone. The ranger tells me their official position is that there are no breeding populations in New Hampshire. Sightings and scat may come from transient individuals or exotic pets released into the forest. But it is probably only a matter of time before mountain lions return to the newly reforested Northeast.

Then suddenly it hits me. What about a coyote? The hair and catlike appearance of the scat had led me down the wrong path. Sanity returns. The scat must from a coyote. I can hear the disappointment on both ends of the line.

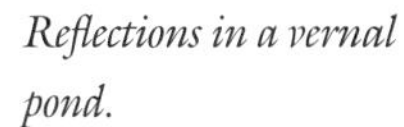

Reflections in a vernal pond.

Chapter 14

Tree Time

APRIL 19, 1999

It is early morning on a Coal Hill farm pond. My daughter and I have hiked up here to check on tadpoles. The sunlight dances and glitters off a curious patch of water nearest the woods. We discover a huge gelatinous mass of eggs dimpling the pond's surface. For the past few weeks, legions of frogs have crawled out of the woods to mate and add their egg clusters to this growing raft of embryos. It is already ten feet across, and growing larger every night.

Thousands of tiny black tadpoles wriggle on the jellylike surface of the raft. They will eat the jelly and lie on its surface to avoid the predators lurking below. We can see newts and dragonfly larvae patrolling the bottom. The occasional swish of a large tail marks the passage of a bullfrog tadpole. A mink frog surfaces through the eggs to glower menacingly at the cavorting tadpoles.

Scattered on the grass surrounding the pond are the desiccated bodies of frogs that didn't make it. Perhaps they were caught by the unseasonably warm dry weather that has made the last few weeks seem more like August than early May.

A northern harrier soars low over the trees, squawks at our presence, then disappears. He draws our attention to something wet and shiny crawling through the grass. It is a spotted salamander that was too preoccupied

Masses of eggs dimple the pond's surface.

A frog that didn't make it.

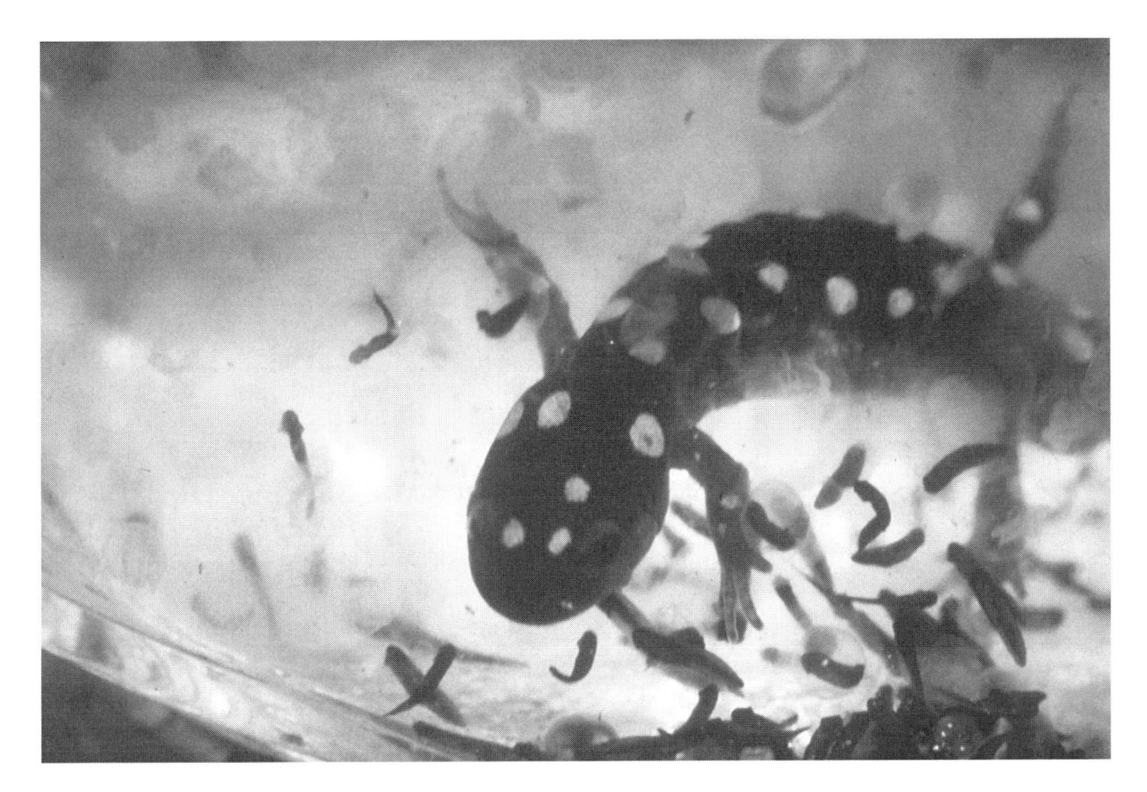

Spotted salamander in tadpole eggs.

with last night's mating to heed the sunrise. Now she is in danger from both the sun and predators. We take some photographs, then carry her to the edge of the forest, where she slithers thankfully beneath the cool leaf litter. There, she will be safe for another night of courtship.

We can still see the tracks of three moose that waded through the shallows of the pond last evening. They were searching for spring's first aquatic vegetation. We follow the tracks out of the pond and into the woods. Almost instantly we are in the moose yard.

In the winter, the yard was covered with patches of ice amidst the snow. Each patch of ice marked where a moose had slept the night before. Their body heat was enough to melt the snow, which refroze as ice the following morning.

But now the moose yard looks like a war zone. The moose have browsed the twigs off the conifers and used their lower teeth to strip the bark from the maple trees. Some of the crisscrossed scars from browsing are eight feet off the ground. Beetles lie in the scars feeding off the sweet sap that still flows from the open wounds.

We push further into the feeding territory. It is a tangle of skinny, straight saplings eight to twelve feet high. Many of the trees have been broken in two and all have been heavily browsed. In former times, farmers would have called this area a sproutland and cut it every ten years for fence rails.

A ruffed grouse picks its way through the saplings, pausing to investigate the orderly piles of moose droppings. Most of the scat are still inch-long, deerlike winter pellets, but here and there are more cowlike plops. The plops indicate that the moose are switching from browsing on trees to grazing on summertime's succulent water vegetation. Just in time too, judging from the state of the overbrowsed saplings.

Suddenly there is a flurry of wings; a small bird flies ahead of us, lands, and starts to probe the soft earth with its long, sensitive bill. This is a woodcock. Soon the evening air will be full of their mating flights. The males fly almost a hundred feet into the air, then tumble back to earth uttering their high, twittering courtship calls.

It seems like all these animals have something to say to us, if I can only figure it out. Woodcock and quail used to be more common here, moose and deer extremely rare.

We sit on a old pine stump and try to figure it out. These are the same pines that confused me so last winter. What went on up here?

Most of the stumps are covered with small, upright crimson growths. They look like tiny toy soldiers, hence their name, British soldier lichen. Thoreau was one of the first naturalists to observe that they are indicators of secondary growth.

We drink in the warm breath of the spring forest and listen to the flute-like calls of a wood thrush. But if we sat here 250 years ago, we would not have heard wood thrush nor been surrounded by whispering pines. We would have looked out over fields, pastures, and orchards. We would have heard plows turning over sod, bobolinks in the meadows, and the bells of distant churches.

The moose yard looks like a war zone.

Moose scat. In the spring it changes from its dry form to its summertime consistency.

The only woods we would have seen would have been at the very tops of mountains or in private woodlots. Every farm needed at least ten acres of woodlot. That was just large enough to keep one stove burning all year for cooking and a another fire burning all winter for heat.

The same thing was true throughout most of New England, where 80 percent of the forests had been removed. Humans engineered the land to suit their needs. Farmers cleared fields to grow grain and pastures to feed cattle. They maintained woodlots for fuel and planted orchards for food. They built sawmills for lumber, iron works for metals, and towns for commerce. Franconia had all these amenities in one valley. But this was a tamed land, operating at close to its maximum carrying capacity.

What happened? Many things. The Civil War introduced New England farmers to other parts of the nation. The McCormack Reaper made it possible to farm the fertile lands of the Midwest. Railroads made it economical to ship grain back to the East for sale. Country life lost its allure and gold was discovered in California.

The result? Hundreds of thousands of people left northern New England. Thousands of towns lost over half their younger populations to the exodus.

An anthropologist from Mars would have said that the civilization of northern New England collapsed like that of the Mayas or Incas. But New Englanders experienced the changes as caused by hundreds of thousands decisions made by thousands of individuals over decades of time.

By 1860, the forests were starting to reclaim New England, but it was a haphazard, piece-by-piece process. Year after year, the older people who stayed behind made small individual decisions. A farmer decided it was no longer profitable to grow grain, so he abandoned a field. His neighbor had lost his son in the Civil War, so he decided to slaughter some cattle and let one of his pastures revert to woodland. Another farmer bought a coal-burning stove, so he stopped cutting his woodlot so heavily.

In Franconia, the Iron Works and sawmill were shut down, so fewer trees were cut for potash, charcoal, and lumber. Yes, the forests were coming back, but the results were not all the same. Which trees grew depended on what the land had been used for before. Plowed fields tended to grow hardwoods, abandoned pastures grew pine trees, and woodlots produced mixed conifers. All could be considered pioneer species, but would they all converge eventually into the same climax forest?

The stump that we are sitting on provides a clue. It is from a pine tree, *Pinus strobus*. When the first settlers came to New England, they were delighted to find huge, old pine trees like these scattered through the woods of New England. Europe had cut down most of its old, large trees centuries before.

Many of the American pines were over two hundred years old and up to 150 feet tall; perfect for the masts of warships. And what country seeking empire wouldn't pay dearly for warship masts? The British decided that pine trees were simply far too valuable to leave in the hands of a bunch of rowdy colonists. They declared that all pine trees over twenty-four inches in diameter belonged to the king, and sent royal agents into the forests to mark such trees with the king's mark, a pair of chevrons chopped deeply into the bark of the unfortunate tree.

Hundreds of thousands of farms were abandoned after the Civil War. Courtesy *The Courier* newspaper.

Pinus strobus.

This edict did not sit well with Yankee woodsmen, who had become accustomed to selling their mast trees to whoever would pay top dollar. This was normally the Spanish, French, and Portuguese—never the British. More than a few of the king's representatives were set upon by woodsmen dressed up as Indians. It became a favorite activity, the North Woods equivalent of a Boston Tea Party. Pine trees became such a cause célèbre that the militia who defended Bunker Hill fought under a flag emblazoned with a pine tree. Vermont still has a pine tree on its state seal, and Maine is "The Pine Tree State" with the pine tree as its state flower.

After the Revolution, settlers quickly took up the Herculean task of clearing the land for agriculture. They did it so well that by the 1800s, 80 percent of the forests had been cleared, along with virtually all of the large, old pine trees.

But after farmers abandoned their fields in the 1860s, the pine trees started to grow. In fact, this was a boom time for pine trees. They had a distinct advantage over other trees because their large seeds could withstand drying and would sprout in abandoned fields covered with grass. Large tracts of pine trees grew on abandoned fields and burned-over areas such as those left on Coal Hill after the charcoal industry collapsed in 1857.

By 1900, these new pines were fifty years old and ready to harvest again. This initiated a second logging boom. Huge areas were clearcut, topsoil was washed away, and erosion fouled rivers and streams. People started to notice and call for change.

Pine in rain.

By 1930, the first environmental movement had started in response to the indiscriminate logging of pine trees. But the movement quickly fell into a squabble between preservationists and conservationists. The preservationists wanted to preserve the land for wilderness and the conservationists wanted to conserve the land for things like hunting and logging.

Theodore Roosevelt avoided this political pickle by creating both the National Park Service and the National Forest Service. The National Park Service was established to preserve natural areas and the National Forest Service was created to conserve forests for their products. In other words, logging has been traditionally permitted in National Forests, including the White Mountain National Forest that surrounds Franconia Notch and includes eight hundred thousand acres of woods and forty-eight mountains over four thousand feet tall.

Today, debates still rage between the preservationist and conservationist wings of the environmental movement. In 1999, the preservationists convinced President Clinton to declare a moratorium on building more logging roads in National Forests. This will make it almost impossible to continue logging in the White Mountain National Forest. An intriguing proposal is also being aired to turn the White Mountain National Forest into a national park, which would ban logging outright.

But this story has another interesting aspect. Along with debates between conservationists and preservationists, there were debates between scientists. Some botanists thought that forests would succeed naturally until a final climax forest is attained, other botanists thought that the final forest would depend on what went before.

I had a relative who was on the wrong side of this debate. He was hired by President Roosevelt to survey the forests of North America, and decided that most of New England had formerly been a pine forest and that pine trees were the desired climax species.

But pine forests do not beget pine trees. Young pine trees cannot grow in the shade of old pine trees. So, the trees that sprout below pine trees are hardwood saplings like those surrounding us in the feeding territory of the moose. These are the trees that will succeed after the pioneering pine trees have been cut away.

The result of this misunderstanding was that foresters spent a lot of time trying to plant pine trees where hardwoods wanted to grow. Predictably, their efforts failed. Gradually it dawned on them that large tracts of pine trees are accidents of history.

The corollary to that lesson is that once we cut down the remaining tracts of pines such as these on Coal Hill, we will not be able to replace them unless we are willing to undertake, then abandon again, the farming practices of the 1800s.

So now I think I can start to understand what happened on this hill. Two hundred and fifty years ago, this area had been denuded, burned by the charcoal industry, then abandoned. White pines had grown up in the burned-over area, then been cut during the logging boom in 1900. Some had been logged again around 1950, and the hardwood saplings had started to grow. Today, these saplings provide browse for moose and deer. Unbelievably enough, forest animals such as moose, deer, bear, turkey, beaver, wolves, and mountain lion had been almost totally extirpated from New England in the 1800s.

But what happened to the animals that were common to the open areas of the early 1800s? The meadowlarks, bobolinks, and bluebirds that used

Pine trees on Coal Hill.

Deer were extirpated in New Hampshire during the 1850s!

Has economics dictated that recreation is the best future for New Hampshire? Courtesy *The Courier* newspaper.

to sing on the fence rails of open fields have been replaced by wood thrush. The woodcock and quail that used to feed in pastures have been replaced by turkeys and grouse. Moose, bear, deer, and beaver have all returned with the new forests. Coyotes have replaced wolves and mountain lion as the top predators in the new forests, but many expect the old predators to return.

The pace has picked up in the last thirty years as the new forests have matured. Today, it is common to read about bear invading rural backyards, moose wandering into cities, and beaver flooding suburban homes. Such things would have shocked the good citizens of the 1800s. The largest wild animal they were accustomed to seeing were muskrats!

New England is the largest area of the world that has undergone such extensive farming, then been allowed to revegetate. But it happened so slowly and subtly that it is only in the past few decades that we have recognized the fact. Today we have only the faintest collective memory of what New England looked like during its agricultural heyday. In the same way, when we look at the jungle-covered ruins of ancient Central American cities, it is difficult to imagine the hundreds of thousands of acres of open agricultural land that it took to support those civilizations.

If you look at a map of New England in the early 1800s, you will notice something else. Human habitation was spread out evenly in small agricultural communities. In 1810, the three largest towns in New England were Boston, Salem, and Nantucket.

But if you look at a map of New England today, you will see almost no open areas. Eighty percent of New England is forested and only twenty percent is open, exactly the reverse of the situation in the 1800s.

The first map looks like a photomicrograph of healthy tissue with cells spread out evenly, while the second map looks like a photomicrograph

Reflections.

of tissue with cancerous tumors. Big black blobs of urban sprawl have sprouted arteries to draw energy and resources from outlying areas. The diagnosis is unmistakable. The first map looks self-sustaining and healthy. The second map looks short-lived and lethal. The tumors may be thriving, but they are short-circuiting the life of their host.

Of course, maps are just metaphors. Another way of interpreting these maps is that humans, abetted by ecology and the dubious guide of economics, are in the ongoing process of re-engineering the world to make it more productive. The invisible hands of both ecology and economics have decided that the Midwest is the best place to grow grain, the Far West is the best place to grow cattle, and the coasts are the best place to grow people.

Where does that leave New England? Has economics dictated that northern New England is the best place to grow trees for scenery, snow for recreation, and parks to nurture city folk? Will economics dictate that New Hampshire, Maine, and Vermont become resort communities with all the economic dependence that that term entails?

The days of self-sustaining agriculture are irretrievably gone. The days of logging are probably going fast. It may be demeaning to be designated a resort area, but isn't providing recreation to millions of bodies, nurture to millions of souls, and oxygen to the world a noble calling as well?

Having figured out the history of Coal Hill and charted the future of New England, we decide to call it a day. The sun is warm on our backs and there is optimism in our hearts as we walk back home for lunch.

The serious demeanor of Osiris, Egyptian God of the underworld; the peregrine falcon.
Courtesy *The Courier* newspaper.

Chapter 15

Beneath Falcons' Wings

> *The peregrine falcon is perhaps the most highly specialized and superlatively well developed flying organism on our planet today.*
>
> —Gerald Henderson Thayer, *Bird Lore*, 1904

JUNE 1, 1999

It is early morning; hot, hazy, sultry as mid-summer in Venice. A peregrine falcon, *Falco peregrinus anatum*, flies in high circles a thousand feet above Franconia Notch.

Deep, dark, dispassionate eyes survey the forests below. Black feathers cloak the peregrine's head, giving him the hooded appearance of the king's executioner, not really concerned about the fate of his victims but supremely confident in his use of the axe. Osiris, the Egyptian god of the underworld, had the same serious demeanor as a peregrine falcon. Ancestors of this bird may have flown with the Great Khan of China, who hunted with ten thousand falconers while sitting on his throne carried on the back of four elephants.

Now a flock of mourning dove flies from a trailhead meadow. The peregrine sweeps lower. He locks onto a single bird flying imperceptibly slower than its flockmates. The valley goes suddenly silent. The peregrine folds in his swept-back wings and starts his power dive.

The dove tries to make it to the deep woods. If she can plunge through the tangle of branches, she might be able to foil the peregrine's attack. Just before reaching the dove, the peregrine stretches out his talons and slams into her back at almost two hundred miles per hour. The dove's body falls limp, feathers fly, and the thump of their impact reverberates through the valley. The peregrine tumbles below the dead bird, rolls up, and catches it in mid-air.

A thousand feet above, the female falcon calls excitedly to her mate. She has watched the kill from their guano-splattered ledge a hundred feet down the vertical face of Eagle Cliff. The male peregrine lands nearby, plucks the feathers off the breast of the dove, and presents it gently to his mate. She accepts the offer, takes a few bites, and feeds the rest to their ravenous chicks.

The female falcon defends her nest. Audubon Society of New Hampshire; photo by Chris Martin.

The view from Eagle Cliff, a spot which has probably been a magnet for falcons since the glaciers retreated through the valley twelve thousand years ago. Courtesy *The Courier* newspaper.

Their nest is on Eagle Cliff, perhaps the most famous peregrine ledge in all of North America. The cliff has all the prerequisites of a perfect nesting site: a vertical cliff impossible for weasels and fox to climb; a ledge that commands a wide valley and lake; river, field, and forest habitats for their food species. The cliff has probably been a magnet for peregrine falcons since the glaciers retreated through the valley twelve thousand years ago.

The ancestors of these birds probably perched on this ledge to prey on the millions of passenger pigeons that once migrated through this valley. The pigeons may have helped reforest the valley by seeding and fertilizing the glacial landscape with their droppings. Later, the sheer weight of thousands of pigeons alighting at once was often enough to break huge branches off trees. The birds were easy pickings for peregrines.

Eagle Cliff is named for a pair of golden eagles that used to nest on this same ledge. The eagles moved on when the forests started to encroach on the open lands below. Today, the ski slopes of Cannon Mountain and the

Their ledge commands the lakes, forests, and open land habitats of Franconia Notch. Audubon Society of New Hampshire; photo by Chris Martin.

trailhead meadows provide the only open areas to attract passerines and gamebirds. Echo and Profile Lakes provide habitat for shorebirds and the forests provide habitat for woodland species; all food for the dominating falcons.

During the early 1900s, oologists would scale these ledges to collect peregrine eggs for their private collections and climbers would rappel down the cliffs to capture chicks for falconry. But the peregrines kept returning to the cliff because it was too perfect to abandon. If a mate died, it was quickly replaced. If a pair failed to return, another pair would take over the site.

But then catastrophe struck. Eagles, falcons, and ospreys disappeared from former nesting sites across the planet. The culprit was DDT. The "wonder pesticide" had helped win World War II by saving allied soldiers from disease-carrying insects, but its widespread use after the war led to unforeseen consequences.

Toxins from DDT's breakdown worked their way up through food-chains from mosquitoes to songbirds; from aquatic creatures, to fish, to shorebirds; from fields, to pests, to gamebirds. And, at the top of each of these pyramids were raptor predators such as eagles, osprey, and peregrine falcon.

The toxins magnified in raptors' tissues were sometimes a million times higher than in the environment. Many peregrines died outright, all started to lay thin-shelled eggs. Some parents ate their eggs in their confusion, others crushed their unhatched chicks in brooding.

By 1964, no peregrine falcons nested in all of eastern North America. The ledges of Franconia Notch were devoid of peregrine falcon for the first time in ten thousand years. The situation was similar on every continent of the world except Antarctica. Peregrine falcon had become the poster children of the environmental movement.

Yet, little by little, things changed. Rachel Carson galvanized the environmental movement with her landmark book, *Silent Spring*. Canada banned DDT in 1971, and the United States followed suit in 1972. Ornithologists convinced the federal government that peregrines could be raised in captivity and released in the wild. They had always been the favorite bird for falconry because they were so easy to train. Critics scoffed at the idea.

But in 1981, something wonderful happened. A wild peregrine that had somehow survived the depredations of DDT landed on Eagle Cliff and took over a raven's nest.

A few days later, he saw a mere speck of a bird flying up the valley. She was just a tiny black cross flying against an immensity of white cumulus clouds, yet something in her wingswept form triggered an innate recognition pattern, though he had not seen another member of his species for several years.

The peregrine launched himself off the ledge and proceeded to fly back and forth across the face of the cliff, uttering excited creaky calls. The female seemed curious but cautious. She too had not seen another peregrine since her release from a special hacking box in coastal New Jersey. She rose above

Something in her wingswept form triggered an innate recognition pattern. Audubon Society of New Hampshire; photo by Mike Pelchat.

the cliff and cut wide circles over its brow while watching the actions of the frantic male below.

The male dashed from ledge to ledge, as if extolling the virtues of each. The female flew nearer to investigate. The male bowed his head to the ground. The female alighted and did the same. The male crouched into the raven's nest, continuing his whining creaky calls. She watched him slowly rotate in the nest as if preparing it for eggs. Something about this ledge and this small wild male pleased the falcon. She decided to spend the night.

The following day, the male peregrine flew out of their scrape and soared over the valley in wide, high circles. The female joined him. Together they filled the valley with their rapturous calls. All spring they could be seen tumbling out of the sky, locking their talons in mid-flight, touching bills at sixty miles an hour. The male started capturing small birds for the female and passing them to her in midair. Often they hunted in tandem. The smaller male would rake the vegetation to scare up large birds for the female to capture. Between them they could kill almost any bird in the valley.

One day, after a particularly energetic aerial display, the female lowered her body onto the ledge and called to her sky-dancing mate. He approached the ledge with a slow exaggerated flight, bunched his talons, and landed on her back. Three weeks later, the female laid the first of their eggs.

By the end of July, the pair had raised two chicks. They were the first wild chicks reared in eastern North America since the pre-DDT era. The ledge has been occupied ever since. One of the pair's chicks migrated to New York City, where she was nicknamed "Queen" and became quite the celebrity.

A climber rappels down the face of the cliff.

The female falcon attacks a climber. Audubon Society of New Hampshire; photo by Mike Pelchat.

She nested for several years on the Throggs Neck Bridge, producing numerous chicks. Evidently she found the rivers of Manhattan, the ledges of skyscrapers, and the pigeons of Central Park suitable substitutes for the amenities of Franconia Notch—there is just no accounting for taste.

But now the sun is getting higher in the sky. A pair of climbers appears on the spire at the top of the cliff. The female flies off her perch and circles threateningly overhead. One of the climbers takes out binoculars and peers at his compatriots in a parking lot three-quarters of a mile below.

"Either they see us or they're swatting at blackflies."

A car pulls out of a space and pauses briefly in the center of the lot.

"That's the signal. We must be right over them."

The first climber jumps off the top of the cliff and rappels down the vertical face. His foot dislodges a boulder, which bounces off the ledge where the female perched only moments before. Was that what happened to her mate, leaving her alone to raise their chicks? She dives at the climber, raking his helmet with her talons.

"Be glad you're wearing your brain bucket. She left a huge gash on top of it."

"Well, if something happens to me, I want to go down the mountain in a Stokes litter, not in one of those heavy pieces of crap."

The second climber laughs briefly, but then is all business. He reaches below the overhang and grabs two chicks. The enraged birds peck at the climber's hands as he affixes metal bands to their legs and gently checks their ears for parasites.

A thousand feet below, the group of volunteers from the New Hampshire Audubon Society stop swatting flies long enough to congratulate themselves. It has been a successful year. These are the twenty-first and twenty-second chicks raised on this cliff in the past sixteen years. This has been one of the most productive cliffs in all of North America—sacred ground in ornithological circles.

But now the sun is setting and the peregrines have returned to their ledge to settle in. The female is still upset with the male. He was hunting on the other side of the valley the whole time the climbers were handling their chicks.

A thousand feet below, a hiker scares a grouse into the air. She flies for cover but it is too late. The male plummets toward the valley. He rakes the vegetation over the clearing where the grouse landed, while the female soars overhead. Finally, the grouse can stand it no longer. She breaks from cover and never sees the female peregrine that knocks her lifeless.

The grouse is several grams heavier than the female. She grasps its dead weight into her talons and slowly flies up to their ledge. The grouse will provide several days food for her hungry chicks. She seems to grudgingly accept that she could not have made the kill without her partner.

By sunset it is over. Only a pile of feathers and the naked feet of the grouse remain. On the parking lot below, I fold up my observing scope and get into my car that served so well as a makeshift semaphore when the ornithologists' walkie-talkies failed. Had we been poachers, we could have

Audubon scientists prepare to band this year's peregrine chicks. Audubon Society of New Hampshire; photo by Mike Pelchat.

The twenty-first and twenty-second peregrine falcons raised on Eagle Cliff. Audubon Society of New Hampshire; photo by Chris Martin.

sold the peregrines for $10,000 to falconers in Asia. Had we been caught, we could have been fined $20,000.

Peregrines are by no means out of the woods. There are still only fifty breeding pairs in all of eastern North America. But I had been privileged to witness one of the most joyous success stories in all of nature. Personally, I'm glad it was in Franconia Notch, not in downtown Manhattan—then again, there is just no accounting for taste.

A deer steps out of the forest.

Chapter 16

Mount Washington

JUNE 12, 1999

It is early morning. I'm sitting on the dew-covered grass watching the meadow fill with golden light. Copses of cinnamon fern shine with a buttery green luminescence. The brook gurgles happily beside me.

A deer steps out of the forest and seems to glow in the early morning light. She has a gentle curiosity about her. Her large black eyes stare deeply into my own. Her velvety ears are alert to the slightest noise. We watch one another for twenty minutes. I finish one film, change rolls, and shoot some more. She stretches, humps up her back, and seems to cough. Bulges ripple down her sides. Eventually it dawns on me that she is supremely pregnant and probably in labor. I let her saunter back into the forest to give birth in private.

Normally I would be enjoying my coffee on the porch, but I have been displaced. A pair of phoebes have built a nest in the eves and insist on their privacy as well. I'm beginning to feel left out.

A junco lands on the red pine beside our house and a puff of yellowish-green pollen drifts into my coffee. A gust of wind produces a downdraft of pollen that floats through the woods like a cloud of slowly moving mustard gas.

It is going to be another hot muggy day, although it is only early June. A week ago it was ninety-seven degrees in Boston. Perhaps Mount Wash-

She seems to glow in the morning light.

ington will provide some relief. There are still snowfields in Tuckerman's Ravine. Skiers climbed to the headwall over Memorial Day Weekend, though it was sixty-four degrees on the summit. Still, Mount Washington should be cooler than this sweltering valley.

I decide to take two detours. The first is to Robert Frost's old home in Franconia. The lupines are in full bloom. Robert (I've always wanted to call the great man Bob but never had the gumption) used to admire the lupines from his front porch while looking at the mountains. Today the lupines remain, but the mountains are disappearing behind sixty-foot pine trees. They have grown in the seventy short years since Frost lived here. It is another reminder of how quickly the forests are reclaiming New England's farmlands.

Eventually it dawns on me, she is supremely pregnant and in labor.

Reforestation in one lifetime. Robert Frost used to sit on this porch and admire the mountains. Today they are hidden behind seventy-year-old pine trees.

My second detour is to Moat Mountain, twenty miles south of Mount Washington. Moat Mountain has a singular place in geological history. It is where moat ring-dike complexes were first described.

Two hundred million years ago, our planet consisted of one globe-girdling ocean, Panthallasia, and one supercontinent, Pangea. But then something remarkable happened. A huge plume of molten magma broke from the earth's mantle and started to melt its way toward the surface. It separated below Pangea, and convection currents started to pull the two sides of the continent apart. Five-hundred-foot fountains of lava spurted out of fissures hundreds of miles long. Massive rivers of viscous magma oozed out of the ground and paved the countryside beneath successive flows.

Eventually, an area equal to the size of Australia was covered with a blanket of volcanic materials over a mile thick. The continent rifted apart and Pantallasia rushed in. The Atlantic Ocean was born. It was probably the greatest amount of volcanism ever to engulf our planet.

In places, vents opened up around huge blocks of volcanic material, and the blocks collapsed into the underlying magma chambers. The result was a moatlike ring of dikes surrounding a circular block of volcanics. Today, the dikes help protect the volcanics from erosion. At one time, all of New Hampshire was covered with volcanics nine thousand feet thick. Now they can only be seen in moat dike structures such as Moat Mountain and Mount Ossipee, to the south. The rest have been eroded away.

The same extensive lava flows built up the Palisades of New York and similar structures in Europe, Africa, and Brazil. They mark where the original continent was pulled apart. It was also during this time that plumes of magma oozed through the bedrock but solidified underground. Today they make up the granite plutons of the White Mountains.

Scientists have recently discovered that all this volcanism caused another great change. Carbon dioxide released from the volcanoes warmed the climate and altered the atmosphere. Only those plants and trees able to withstand carbon dioxide survived, and the meat-eating dinosaurs died off, preparing the way for true dinosaurs.

As I drive north from Moat Mountain I'm driving into a much older domain. Remember those sands and muds that lay on the bottom of the Iapetus Sea during the Ordovician? Four hundred million years ago, they were being buried and squeezed under seven miles of bedrock into quartzite and schist as the Iapetus Sea closed. The earth's plates were ramming together, thrusting up mountains as high and spectacular as today's Himalayas. This was the Taconic orogeny. Later, under even greater pressure, the quartz and micas separated into black-and-white-banded gneiss (pronounced "nice").

So the Granite State's most revered mountains are not made up of granite at all, but of gneisses and schists. These names are seared into the hippocampus of anyone who has ever taken "Rocks for Jocks," the introductory geology course at Harvard College. Gneisses and all those other schists made such memorable punchlines to the limericks we concocted to memorize the geological ages. Granted the limericks were sophomoric—but so were we at the time.

I'm told that today the geology department crams a lot of petrology into "Rocks for Jocks" in order to enhance its reputation. But it will be a losing game; such nicknames survive long after their owners have gained maturity. Besides, it was one of the best courses I ever attended.

Mount Washington has had a long association with New England col-

Lupines.

Sunset on Bigelow's Lawn, Mount Washington.

Felzenmeer, "a sea of rocks."

leges. During the nineteenth century, the summit was a favorite site for elegant summer field trips. Distinguished scientists would ride stage coaches to the summit and spend the night at the Tip Top House. I like to think of Agassiz, Bigelow, and Tuckerman sitting under umbrellas divvying up the nomenclature of New England. How did Tuckerman get a ravine, Bigelow get a lawn, while Agassiz only got a rock?

After the cog railway was completed in 1869, more scientists arrived. The entomologist Annie T. Slosson describes a group of botanists, "flying from side to side of the train car, looking eagerly out and uttering strange exclamations, such as, 'Geum!', 'Ledum!' . . . springing from the train at its brief stops to collect plants to the intense amazement and amusement of the unscientific passengers."

You still see people exhibiting the same mildly eccentric behavior alongside the modern auto road. It seems to keep them young, enthusiastic, and fascinated with nature—and still bemuses their neighbors.

Outside my car, the hardwood trees have slowly given way to spruce then balsam fir. As I labor past three thousand feet, the fir start to become stunted and impenetrable, a solid barrier of "tuckamore," probably the only truly never-cut forest in New England. Finally, I reach the crooked wood or krummholz. Here, balsam fir look like flags or broomsticks, while black spruce grow as ground-hugging mats. At 4,300 feet, the road bursts through the timberline and skirts around a hairpin turn called the Horn. I am surrounded by a sea of jagged ice-shattered boulders called felzenmeer.

The timberline is determined not so much by the cold or wind, but primarily by the lack of summer heat. The length of the summer growing season determines whether a tree has enough time to produce new growth, so the treeline in both the arctic and alpine environments is close to where it averages 54 degrees Fahrenheit during July, the warmest month of the

Agile as a mountain goat and barely breathing.

A cairn topped with a capstone of white quartz.

year. One way to gauge global warming would be to measure how quickly treelines are migrating both up in the mountains and north in the arctic. We have seen as much as twenty-mile migrations in treelines in southern latitudes in recent years.

Just before the seven-mile post, I pull into a parking lot. A climber jogs up Huntington Ravine Trail, crampons and carabiners jangling off his tattooed thighs. He is agile as a mountain goat and barely breathing. Two botanists come puffing behind.

A cairn draws me down the trail. It sparkles with bits of mica and is topped with a capstone of white quartz. Dark green crystals of sillimanite shimmer in the heat. I turn right onto the Alpine Garden Trail. A snowbank nestles into the felzenmeer above me. Below it spreads a dark green beard of vegetation. There are patches of krummholz, hair grasses, and sedge.

I cross the tiny rivulet that drains off the snowpack. Succulent green leaves of mountain aven line the bank of the Lilliputian river and I almost expect to see little people fishing from its banks. Tiny stalks of elfin flowers rise above pincushions of deep green evergreen leaves. These are the deli-

A patch of snow nestles into the felzenmeer.

cate white and yellow flowers of diapensia that grow on the coldest, most windswept crags of the mountain. Their dark green leaves and pincushion form help them retain heat and water.

It is difficult to imagine that I'm on the same mountain that has claimed 124 lives and has hurricane-force winds and and freezing weather every month of the year. Today it is 66 degrees, but feels much hotter. All I can hear is the sound of my heart and the summer sounds of an amazing profusion of insects. A swallowtail butterfly sips nectar from magenta blossoms of Lapland rosebay, while bees and bee flies hover over the delicate pink flowers of alpine azaleas. Every crevice seem to house another bonsai garden of perfect, zenlike flowers.

Diapensia grow on the most windswept crags of the mountain.

Sandwort.

I am totally alone on this six-thousand-foot peak. There is no wind, no sounds save for the quiet buzzing of these bees. Nothing is quite so summerlike nor ethereal as an alpine meadow in June.

A deep black wolf spider skitters over the boulders and I hear something faintly. I clamber up the trail and the sound becomes louder. Could it be, a polka? On Mount Washington?

Suddenly, the perpetrator comes roaring up the mountain. He is a middle-aged biker with a grin as wide as the road and his radio on full blast. He has the look of someone who has been busily blowing out brain cells for most of his life but has kept his pleasure center firmly intact, his polka

"Just Married."

The summit.
© Mount Washington Observatory Photo.

gene in full flower. I suppose I should feel intruded upon but I don't. The polka is jaunty, kitschy, and somehow fitting.

Throughout the day, I continue to see a greater diversity of people than I would in any city park. They range from incredibly fit climbers to naturalists intent on identifying the binomial nomenclature of every species on the mountain. A Florida couple has driven up here just so they can buy a few souvenirs, display a "This Car Climbed Mount Washington" bumper sticker, and call some friends from the summit of this mountain as spiritual as Mount Fuji and as funky and American as Disneyland.

PART IV

Summer

We fancy that industry supports us, forgetting what supports industry.

—Aldo Leopold, *A Sand County Almanac*, 1949

Wild lupine.

Mirror Lake.

Chapter 17

Hubbard Brook

JULY 6, 1999

Four o'clock a.m., July 6—The fan is going full blast, but it is still too hot for sleep. I am lying on top of the sheets, but the muggy air covers me like a steaming wet towel. Mirror Lake is silent. No breeze blows across its surface. I am staying at the Hubbard Brook Experimental Forest forty miles south of Franconia.

It has been hot and humid for several days—too hazy for photography. Heat advisories have been broadcast from the Mississippi to the East Coast. It is over 100 degrees in Atlantic City, Philadelphia, and Washington, D.C. The heat index is 115 in West Virginia. Thirty-one people have died from heat-induced cardiac collapse in New York City. The city's power went out for eighteen hours, and the water pressure is dangerously low because so many people have illegally opened fire hydrants. Who says global warming is not disrupting our civilization?

The culprit is La Niña, the same global atmospheric pattern that caused the 1988 drought. That was the year that cornfields parched in the Midwest and riverboats were abandoned as the Mississippi ran dry. It was the year that an eighty-year-old farmer broke into tears on national television as he sifted powdery soil through his gnarly dry fingers. That was the year that concern about global warming peaked—the year before newspapers returned to scandals and the booming economy.

Lightning behind the pines.

I am awoken from my hypnogogic state. Constant flashes light up the sky. "Must be the grande finale," I decide in my semi-delusion.

My watch jolts me back into reality. "Four-thirty. Can't be fireworks; must be heat lightning." But it is eerie how it lights up the sky with its frightening hypnotic flashes. "Has to be a storm, but it's too far away to affect us."

I keep my observations to myself. My wife hates my crushingly obvious meteorological commentaries: "It's raining outside. Boy it sure is hot. Yeah, but it's not the heat, it's the stupidity." Such observations don't go down smoothly at 4:30 in the morning.

But now a faint rumbling rolls out of the western sky. A pheasant responds. The rumbling grows louder. The constant flashes resolve themselves into discrete cracks of lightning.

I count the seconds until the next roll of thunder. The storm is three miles away. A faint breeze starts rustling the leaves. A minute later, the storm is two miles away. Now it is starting to get very dark again. I rush around the cabin madly collecting candles, pulling appliance plugs, closing doors and windows.

"This thing may really hit," I yell as I grab my camera and head to the shore. There is just enough light to see the silhouettes of the mountains surrounding Mirror Lake. Tourists used to cross the lake by canoe every Saturday night to attend dances at an outcrop on the far shore. But now the dancehall is gone, along with a sarsaparilla soda bottling works, a tannery, and two sawmills.

A blast of wind flings ripples across the lake and almost knocks me off my feet. Overhead, a mass of cold air is trying to shove itself under the blanket of muggy air that has plagued us for the last three days. The earth seems to be in a titanic struggle to cool itself off. Hot air spirals into towering thunderheads where its humidity is frozen. Then it melts and falls back as large, fat drops of cooling rain. Invisible forks of lightning zig-zag out of the sky. They are pulling electrons to the tops of mountains, trees, and to the surface of the lake.

A huge crack of lightning signals that the negative electrons of the clouds have met the positive electrons of the lake surface. They are now racing toward the cloud at a third the speed of light. The fluctuating bolt is five times hotter than the surface of the sun. It pulsates and flickers for half a second.

If I can only find a boulder large enough to steady my camera, I should be able to get a smashing picture. Then again, I can almost feel the electrons crackling at the top of my head. Another downdraft of wind strips leaves from the trees and stings my face with lashing raindrops. I have just enough time to rush into the cabin and slam shut the door before the lights go out and the deluge descends.

Fifteen minutes later, the storm has moved eastward and the air temperature has plummeted twenty degrees. The hurricane-force downdrafts have snapped huge branches and left the roads strewn with shredded leaves. But we are now firmly in the grip of the welcome cold front.

Overhead, a soft cool rain is falling quietly in the mountains. It fertilizes the forest with nitrates formed by the lightning. It splashes on outcrops of granite and schist. It weathers out molecules of calcium, magnesium,

A blast of wind flings ripples across the lake.

A soft cool rain.

potassium. It deposits nitrogen from car exhausts and sulphur from power stations. It drizzles into the soil and is taken up by the forest's rootlets. Some of the rain escapes into brooks and streams, but enough of it falls onto weather stations so that scientists from the Hubbard Brook Experimental Forest can tease apart its story.

But right now the "Brookers" are unhappy. It is 6:00 a.m. at Pleasant View Farm. Their kitchen stinks of last night's codfish dinner. They have no hot water to wash the dishes, no cold water to fill the toilets, no power to run the labs, no electricity to generate the high-tech computer graphics needed for the annual cooperators' meeting.

"Perhaps we should start moving food into the pit."

"The pit? What's the pit?"

"See that trapdoor in the kitchen floor? It leads to a two-hundred-year-old cellar. That's where Bormann and Likens used to analyze their water samples. They still love to bring dignitaries to the farm and show them the pit. You can see the visitors saying to themselves, 'That's it? That's where all the great discoveries were made?'

"Of course, the National Science Foundation only knew the pit as 'the constant temperature lab.' That's what it was called in the official grant proposals, but it was really just that dirty old cellar."

Things weren't going much better at the National Forest Service Headquarters just up the road. Ian Halm had arrived at 6:00 ready to hike up to the meteorological stations. But he was met at the door by Ralph Terron,

"Before you do anything, you better check on the pump. I think it's air-bound. Chris thinks we may have to cancel the meetings."

"We can't do that! These people are coming in from all over the world. He'll just have to tell them to pee in the woods. Not in any research areas mind you, but south of the building they'll be on their own."

"We'd better check on the met stations. I think we're going to have a lot of work ahead of us."

Indeed, the next week would be like the old days at Hubbard Brook. Then everybody ate, argued, laughed, loved, and sometimes slept together at Pleasant View Farm. Today the research station has modern facilities and a complicated mix of technicians and scientists from different labs and government agencies, but the field work is about the same.

By 6:30, Ian is at the top of the first watershed. He uses a dipstick to measure the rainfall from last night's storm and collects charts that recorded the week's temperatures and humidity changes.

Then he climbs several miles down the watershed to a large concrete weir and stilling pool. Here he collects streamflow data recorded as the brook flows through a V-notched steel plate. Above him, a hydrologist slowly slogs up the mountain, pausing every few minutes to collect water from one of the tiny rivulets that drains the watershed. He carefully labels each plastic bottle and puts it into his bulging backpack.

The bottles will be analyzed and added to the thousands of other water samples that form an uninterrupted record of the chemistry of the watershed that stretches back thirty-six years. It is something of a world treasure, the most complete database of any ecosystem in the world. It has allowed scientists to gain a more profound understanding of how forests really

Instrument station.

Collecting streamflow data.

work, what happens to them when they are altered, and how they fit into global energy budgets.

The technique has also allowed hydrologists to diagnose the health of the forest. They do this by measuring the input and output of the watershed the way a doctor samples his patient. They claim their method is like taking a blood sample, but it is really more akin to a urinalysis—too much bilirubin, a faulty liver; too much creatinine, evidence of a heart attack; hypoosmotic urine, diabetes insipidus; hyperosmotic pee, diabetes mellitus.

It is only marginally different in the forest. Too much water flow, damage from clearcutting; hyperosmotic streams, acid rain; less sulphur in the rain, things are improving; low pH in the streams, things are about the same.

"So, has the Clean Air Act worked or not?"

"I'd have to say that the Clean Air Act cleaned up our air act, but we're not out of the woods yet, with acid rain."

"Sort of like, 'The treatment was a great success, but unfortunately the patient died.' "

"Sort of like that, yeah."

Ecological sampling does that to you. It's slogging slow work that gives you lots of time to make up horrible puns.

"What's it like up here in winter?"

Ian pauses to demonstrate that pineapple weed does indeed smell like pineapples and can be used to make a passable tea.

"Winter? Pretty much the same routine. We still have to maintain the stations and collect the data every week. Sometimes we have to put floating propane burners into the weirs to keep the water running. One year I had to use my snowmobile to haul a rowboat eight miles up a watershed to repair one of the stilling basins.

"Last autumn, I was charged by a bull moose. I had to run behind a yellow birch, and then we had a standoff. I was only four feet from the tips of his antlers. Finally I realized that three calves and two cows were behind me, so I sidled away while he gave me a long significant stare. Several winters ago, I caught a glimpse of a mountain lion. Of course, my friends at Fish and Game don't believe me, but the forest seemed a lot less tame for a few weeks after the encounter."

Hubbard Brook's biggest success came in 1972, when scientists used the watershed databases to document acid rain. It was the first time acid rain had been described in North America. Today, they are using the same water samples to show that although sulphur has been removed from acid rain, soils have lost their ability to buffer water and streams are still acidic.

This fall, Hubbard Brook will use helicopters to broadcast calcium over Watershed One to see if they can "medicate" the forest back to health. A healthy forest acts like a kidney to balance the water budget of the ecosystem, so that valuable nutrients such as calcium remain in the system cycling from through the trees and soil, without being flushed out of the forest in streams. It will be the first time this technique has been attempted and Hubbard Brook will be watched closely by scientists and governments throughout the world.

The geochemists who measure how nutrients flow through Hubbard Brook are "whole ecosystems" people. They helped change ecology from a merely descriptive science to an experimental one. But at Hubbard Brook, they still rub shoulders with old-fashioned "bucket biologists."

Kate MacNeale collecting stream plankton.

Kate MacNeale is such a "single species" ecologist. Right now she is bushwhacking through "beech hell" trying to catch an adult stonefly. She has laced its stream with an isotope in order to identify the adults after they emerge from the water. It will allow her to know what part of the stream they lived in as juveniles.

These primitive insects recapitulate one of evolution's major steps. They spend most of their lives as voracious nymphs living in clear-running, oxygen-rich mountain streams. It is only as adults that they emerge from the water, having developed trachea for breathing air and wings for flight.

Not very good flight as it turns out. This adult has only flown three hundred meters upstream from where it emerged. Most species don't make it that far. The adult lands on the pale stem of a newly emerged Indian pipe and is quickly captured.

Kate pauses in her work, "Some scientists think that wings were first evolved by aquatic insects to help them skim across the surface of water. The wind would push the adults away from where they had emerged, insuring that they would deposit their eggs in a different patch of water.

"Once you had wings for skimming, it was just a short jump to evolve wings for flying. It's something like neoteny, where a creature develops a trait in a different stage of life, then assumes it for its dominant lifestyle. It's a way of speeding up evolution. Some people think humans developed that way. When you think about it, we're really just chimps with infant-sized heads."

On a nearby slope, an ornithologist is bushwhacking through a clearing in search of "pain-in-the-ass," a black-throated blue warbler that has avoided revealing its nest all summer. Now the bird sits on a twig singing a chirpy little territorial song. But it is probably a ploy. He is waiting for a nearby male to fly off in search for food so "pain-in-the-ass" can mate with his neighbor's wife. These birds are unapologetic philanderers.

But the joke is on the male. While he is off stealing a copulation, another male will impregnate his nestmate. The male will respond by spending less time feeding his offspring in order to avoid being fooled into supporting someone else's genes. They do this more as the population rises. It turns out that individual females "know" exactly what they are doing. Multiple mating increases the number of chicks they can produce individually even as it reduces the growth of the population as a whole.

Steve Hamburg has discovered that the warblers feed on snails and that snails are a good indicator of forest damage. Snails are more abundant where there is calcium in the soil and there is more calcium in the soil after a forest has been clearcut. Clearcutting could account for more warblers, making them more polygamous, which in turn diminishes their population growth.

This is the kind of integrated, interdisciplinary work for which Hubbard Brook is famous. Bird by bird, drop by drop, root by root, "Brookers" are slowly piecing together an ever more accurate picture of how an ecosystem works. Their work is being used to understand other ecosystems and the world's nutrient budget as a whole.

"Pain in the ass"; a black-throated blue warbler with chicks. Ted Levin.

"Brookers" realize that no Rosetta stone, no unifying theory will make ecosystems simple and clearly understandable. Ecosystems evolve, and the hallmark of evolution is complexity. It is just this complexity, this redundancy, that may save the world from such simplifying forces as global warming.

At the evening meeting, I ask the head of the Hubbard Brook Foundation whether they have plans to use their databases to study global warming. He indicates that three or four years of warm weather isn't enough data to suggest that global warming has occurred. It is the kind of healthy skepticism that a scientist should properly display in public. But what I want is a more free-wheeling, over-a-beer conversation.

As the day draws to a close I get my chance. The storm is long gone but the electricity is still out. As the darkness falls, graduate students, senior scientists, and technicians wander into the screened-in porch that wraps around the farm. They gather to share a few warm beers and watch the fireflies flit through fragrant sprigs of pineapple weed. A bat flies up and down the porch capturing trapped mosquitoes. I ask again if anyone is concerned about global warming.

"Let me put it to you this way. If I were an ecologist from Mars, I would be struck dumb by your headlines. You writers are missing the most important story of the millennium. Our planet's ecosystems are collapsing one by one, and what do you write about? Who's boinking whom in the White House and how many more gazillion bucks some smart-ass x-er has made on the Internet. Who's gonna care in fifty years when it's too late to do anything about it? Don't you get it? 'Get your heads out of the silicon of virtual reality earthlings. You have a beautiful planet, but you are already so far into global warming that you can't get yourselves out of it.' "

A graduate student chimes in, "Of course, we lap up the juicy bits about our leaders. We're primates fer Christ sakes. We love gossip; it's just evolved social grooming."

"Well, I happen to think that humans are just not equipped to deal with long-term problems like environmental decline. Our brains are hardwired to exploit new territory, populate exponentially, and deal with short-term problems like business cycles and wars. The only reason people's heads are into virtual reality is that it's so much easier to manipulate than the real world."

"Yeah, but it's all based on the real world and some day the real world's gonna rear up and bite us on the butt. Look at the problems we've run into with genetic engineering. We're creating superweeds instead of feeding the world. We had no idea genetically altered crops would hybridize

Praying mantis.

Reflections.

with wild plants. Now they can spead resistance to antibiotics, insecticides, and pesticides."

"Of course, it's you biologists who have sold out." This from a geologist. "You're so damn busy getting new drugs to market and new genes into crops that no one is even thinking about the real carrying capacity of our planet, let alone figuring it out."

"That's because it's such damn, hard, tedious work. Back in the sixties, there was a real sense of mission. The best and the brightest were willing to spend years in the field trying to understand our planet. Today, everybody seems to be giving up. But the problems are a hundred times worse."

"Even Al Gore seems to have given up being the environmental candidate."

"That's why he lost in 1988. You're about as popular as a skunk at a garden party if you talk about the environment when the Dow is soaring over eleven thousand points."

"Well, I still believe that what we're doing up here is the most important thing on earth. It's the only thing that will give politicians the numbers they need to make good policy decisions. We may not make any more great discoveries like acid rain, perhaps ecology is no longer as challenging as tinkering with genes or studying the brain, but I still think it is much more important to the future of the world."

Everyone looked to one of the founders of Hubbard Brook to voice his assent.

"Well, you know, I'm not so sure. All this science and all these facts. They don't really amount to a pinch of coonshit. In the overall realm of things, science won't do much for us unless we develop a profound and really deep respect for nature. I'm just not sure we are going to develop it in time. I know it's a gloomy picture, but the multinationals are in charge. Sure, everybody's got their heads in the sand, but life is still fun. We can glory in the wonder of nature."

It was only later that I realized how much Herb Bormann practiced what he preached. When I first arrived at Hubbard Brook, he asked me if I needed to take a leak and explained he often did so in his garden. When I left, he offered me some homegrown lettuce. It was only after enjoying my dinner that it struck me just how dearly Herb believes in nutrient recycling. I feel that much closer to him for the experience.

Part of the dawn symphony; eastern bluebird. Mass. Wildlife/Bill Byrne.

Chapter 18

Dawn Symphony

> *One had to be versed in country things*
> *Not to believe the phoebes wept.*
> —Robert Frost, "The Need of Being Versed in Country Things"

JULY 7, 1999

"Complicated" has awoken me. It must be 4:30. Every morning he wakes me with this intricate aria of whistles, trills, chips, chirps, and twitters. I think he is a winter wren, *Troglodytes troglodytes*, but I am leery of making any ornithological gaffes. A favorite review of one of my books said,

"Sargent sees best when his head is underwater." I think it was a comment on my knowledge of birds. Most marine biologists separate their birds into big brown birds, little brown birds, and owls . . . Perhaps I should just leave it at that.

I have never seen "complicated bird." He isn't named *Troglodytes*, the cave dweller, for nothing. He will hop about invisibly on the forest floor but his song will dominate the dawn chorus for the next half hour. It will give me time to set up the front porch with the tools of my trade: notebook, camera, a mug of coffee, and a folding chair.

To my right, a large, well-built phoebe's nest sits in the eaves of the porch. All spring, we watched the phoebes rebuild this nest on top of the one they built last year. The male had arrived when there was still snow on the ground in March and was quickly followed by the female. The male would sit on a low branch of a pine tree, bobbing his tail, and darting out to challenge any other flycatcher who tried to invade his aerial territory. The female, meanwhile, did all the heavy lifting. She gathered beakfuls of mud from the banks of the brook and placed them on the nest with a mason's precision. Later she added moss.

But their nesting time is now over. Last year, the pair produced two broods, but this year they produced only one. It has been too dry to support the abundance of insects they need to feed two broods. They, too, may be victims of global warming.

The phoebes are actually the reason I am sitting here. The first summer we started coming to this house, the same pair of phoebes had built their nest on a strut above the back door. The birds not only provided us with hours of entertainment, but they altered our family's behavior as well.

The former owners of the farmhouse had originally used the front door as the main entrance. However, automobiles, under the guise of convenience, had altered the orientation of the house. It was easier to unload cars by the back door, so gradually the graceful slate path that curved up the hill to the front door was abandoned. The lilacs were allowed to grow

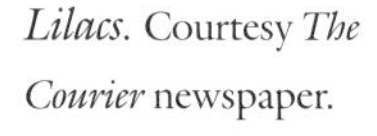

Lilacs. Courtesy *The Courier* newspaper.

Tiger lily.

wild. The front porch, once the place for meals and socializing, was allowed to devolve into simply the place to pile firewood.

The phoebes first convinced us to start using the front door to avoid disturbing their nest, then they chided us into cleaning up the porch. Under their patient tutelage, we removed the firewood, painted the floorboards, trimmed the lilacs, and discovered the slate path that had been gradually buried under grass. It was like unearthing the remains of a former, more graceful civilization beneath your own home. Soon we rediscovered the pleasures of sitting, eating, and reading on the porch. That, in turn, led to the sightings of several deer, a fox, four turkeys, and at least one bear.

It is reassuring to think that a pair of phoebes can still overturn the tyranny of automobiles, and return at least one family to a more fulfilling lifestyle.

To my left hangs the reason we may have to abandon our porch. A colony of white-faced hornets, *Vespula macolata*, have constructed their grapefruit-sized nest below the overhang. They seem to be a lot more active at dawn than at midday. One of them dives into my hair, but exits without stinging. Score one for longish hair. I move my chair out of their sphere of aerial influence.

I discovered the hornets in May, when their nest was barely the size of a plum and would have been easy to remove. Every weekend I threatened to bomb their home, but as Sunday rolled around, they had again proved to be such upstanding citizens that I couldn't face their eradication. Besides, the entire family has become enchanted with our fellow travelers.

My daughter Chappell loves how they drop straight out of the nest and free-fall for several feet before rocketing off under their own power. My wife enjoys watching them chew up wood and add horizontal strips of paper to their nest.

Hairy woodpecker.

It is like watching the construction of a handsome Chinese lantern. Each hornet works on a separate strip of new paper. Some are gray, some are brown, some are an elegant white. They glisten in the early morning sun before they dry and fade. The white strips come from the inner layers of the nest, which they remasticate to make the outer wall, an ingenious form of recycling. We may even experiment with hanging construction paper near their home to see if they incorporate multicolored strips into their design. Could Chappell's Nest become as famous as Charlotte's Web?

By 4:45, the dawn chorus is becoming more crowded. A strident, "Teacher, teacher, TEACHER!" echoes through the forest. It sounds like a schoolboy becoming increasingly alarmed about his need for bathroom privileges. It is the call of the ovenbird, another bird I have never seen. He also spends most of his time turning over leaf litter for food. His nest sits on the forest floor with an opening that looks exactly like the door of a brick oven—if you ever happen to find it.

At 4:46, a robin joins the chorus. It is a sign that the dawn light has reached .01 candlepower.

At 5:01 I hear the first sublime note of the wood thrush. It's flutelike song casts a spell over the forest. It is like hearing a single Gregorian chant echoing through the nave of a vast and empty cathedral.

By 5:15 the dawn chorus is in full song. Fifteen different warblers probably have joined the throng. I can't distinguish their individual songs and lump them lazily together as "little yellow birds."

Noticeably missing from the chorus is the monotonous call of the red-

Swallow-tail butterfly. Courtesy Bill Worthington.

eyed vireo and the "Old Sam Peabody, Peabody, Peabody" of the white-throated sparrow. The season has already turned the corner. Most birds have finished nesting and become noticeably less talkative and territorial.

At 5:28 I hear a new sound. Could it be? Yes it is! A rooster, definitely a rooster, probably on McGowan Hill. The sound carries me back hundreds of years to when I would have heard the lowing of cows eager to be milked and the neighing of horses less eager about bringing in the harvest.

The sharp chattering of a red squirrel wakes me from my reverie. He is soon joined by the distant roar of I-93. Logs are traveling north to sawmills, lumber is traveling south, tourists are arriving for one more day of mid-summer. As the sun heats the land, the lush chorus of bird songs will soon be replaced by the swelling notes of the insect section. Theirs will be bittersweet symphony foreshadowing the coming of autumn.

Heading south.

Turkey. Mass. Wildlife/Bill Byrne.

Now it is 6:00, and I finally see my first bird. It is an unprepossessing robin flying to the top of a pine tree. The sun is now shining on the ferns that have become more cinnamony as the season has progressed. Jewelweeds are starting to poke through the tangle of flowers along the edge of the unmown meadow. A ruby-throated hummingbird has discovered the blossoms and defends them fiercely with strident, bell-like little "tinks." The jewelweed's nectar will provide him with the energy he needs to complete his arduous journey to Central America.

Earlier in the summer, we watched the hummingbirds collect spiders' silk to line their nests. They also used the red color of shelf mushrooms to lead them to yellow birches girdled with scores of holes oozing sap. Yellow-bellied sapsuckers had tapped the holes into the birches, so they were starting to rot and attract the mushrooms. Here is an interesting evolutionary conundrum: Which came first, hummingbirds' attraction to red flowers or the evolution of red pigments in this fungus that lives off birches slowly dying from sapsucker holes?

The meadow below me is crisscrossed with a filigree of dark paths where deer have walked through the new dew. They lead me to a pile of large logs left over from when we rebuilt the barn.

Suddenly, there is a great flurry of wings. I snap a picture reflexively, but my shutter speed is still set for photographing last night's lightning. Even through my lens, I can see that something isn't right. The bird has to be a ruffed grouse, but it is awfully big and gawky.

Seconds later I have another chance. There is no mistaking it this time. We can feel the powerful wings and see the ungainly neck of the mother turkey. She must be upset with her chick for flying. Turkeys usually prefer to stay hidden on the ground and use their legs, size, and vocal chords for protection.

This is a happy sight. Several weeks ago our neighbors heard a ruckus and climbed up Coal Hill to investigate. They found a mother turkey gobbling loudly while running around a young bear. The bear was trying to dig under a fallen tree to get at the turkey's chicks.

The bear eventually gave up, but the next day my neighbors saw a pair of coyotes at the same busy task. Unfortunately coyotes are more persistent than yearling bears. Apparently this chick was the only survivor.

The incident was noteworthy enough to tell a neighbor, but such things are becoming almost commonplace occurrences. Only twenty years ago there would have been no turkeys, bear, or coyotes to play their parts. Once again I am amazed by the rapidity of this forest's recovery. We gather some turkey feathers, place them jauntily in our hats, and hurry home for breakfast. We have a lot to do before our trip to the Palermo mine . . .

Entering the Palermo mine.

Chapter 19

Mica, Beryl, and Schist

The Palermo Mine

JULY 24, 1999

A six-foot crystal of feldspar arches over the small, dark quarry hole that leads to the Palermo mine. It is a fitting welcome to the strange underground world we are about to enter. I inch past a rock hound marveling at the crystal, and start to half-slip, half-slide down the steep incline. Tendrils of icy air weave about our ankles—a welcome relief from the muggy air above.

The light dims as the opening gets narrower. People grope for handholds on walls slick with ground water. We half-fall, half-scramble into the bat room, where a woman with gimpy knees wisely decides to return to the surface to sift through the old tailings of the mine.

Beyond the bat room, we descend through invisible layers of ever cooler air; exhalations of the underworld, they draw us deeper and deeper. Now we are sliding into the belly of the beast, which is reluctant to reveal its secrets. My miner's light shines on walls slick with seep water. We are entering the frozen remains of a tear-shaped blob of magma that melted its way into the overlying country rock 380 million years ago.

The country rock is our old friend Littleton schist, which started as mud on the bottom of the Panthallassic Ocean. It was buried and lithified into shale, then buried some more and heated under the pressure of the overlying rock to become schist.

Then, about 380 million years ago, as the continents were converging overhead, an ovoid blob of magma melted through the schist and began to cool. It cooled from 800 degrees Centigrade down to 50 degrees Centigrade over a long, drawn-out period. It was this long cooling period that allowed the magma to grow large crystals and the coarse-grained minerals of pegmatite. Pegmatite is simply granite that has had a longer time to cool and form larger, coarser grains than granite. But "coarse-grained" hardly seems adequate to describe the formations we are seeing around us, six-foot crystals of feldspar, ten-foot masses of quartz, four-foot sheets of mica.

The books of shiny mica jut out of the ceiling at odd angles. Miners used to pry these out of the walls and load them into horsedrawn carts. The mica was so valuable that it was hauled out secretly at night to avoid being hijacked on the old stagecoach road to Boston. Most of the mica was shipped to England, where it was rifted and trimmed into large translucent sheets to make windows and portholes, for glass was still a rare and expensive luxury.

But mica was just the beginning. The pegmatite didn't all form at once. It crystallized slowly from the outside in. As it did so, the molten magma changed chemical composition and became more liquid as it lost minerals to the earlier forming crystals. This meant that the crystals that formed later had a different composition from the minerals that formed first. This is also

Books of mica.

the reason that the Palermo mine has a greater variety of minerals than almost any spot on earth.

As we continue our descent into the ancient magma chamber, I start to think of it as a huge watermelon standing on end. The outside green rind is where the magma met and partially melted the country rock. It was hottest and cooled the quickest. This is where you find smaller bodies of feldspar, quartz, and biotite mica. The inner white rind is where you find crystals of greenish-blue tourmaline and large sheets of muscovite mica. The pink center of the watermelon represents the forty-foot core of the chamber, where you find crystal-filled pods of triphylite and massive formations of quartz. I suppose the triphylite pods must represent the pits of my metaphorical watermelon.

Mica provides an example of how the process works. We find small plates of black mica on the outer portions of the chamber and large plates of silvery clear mica nearer the core. The black mica is biotite. It formed first, when the magma was hotter and contained iron and manganese, which give biotite its darker color. The clear mica is muscovite, which formed when the magma was cooler and its darker minerals had already been spoken for.

Finally, you find quartzy veins that weave through the formations like developing neurons. The veins formed after the magma was already cool and watery. It seeped into fracture zones and attacked the triphylite pods. The magma altered the chemistry of the pods, making ever-more exotic minerals such as uranium and the lustrous green crystals of palermoite found nowhere else on earth.

Treasures.

Rock hounds.

In all, over a hundred minerals have been excavated from the Palermo mine; eleven of them were first discovered here and four of them are found nowhere else on earth. The United States Army still keeps Palermo on its list of mines containing strategic minerals such as tripolite, beryllium, and radioactive columbianite. Mica is now used to make artificial snow.

It is no wonder that the area has been active since 1863 when Charles Kellogg first started mining the quarry by pouring water into fractures and letting it freeze. The freezing and thawing of the ice cracked off angular hunks of rock, revealing large silvery sheets of mica. At its peak, eighty-five miners worked in the mine with hand drills, power drills, and eventually, more-damaging dynamite.

The mine went through a series of owners, including General Electric, which mined it for feldspar and beryl; and Harvard College, which financed research on the exotic minerals of the triphylite pods. Today, Robert Whitemore operates the mine for mineralogists such as this group that I have joined from Harvard's Museum of Natural History.

The mine is still a treasure that yields amateurs striking slabs of mica, exquisite samples of beryl, and crystal-filled pods of triphylite. None of the minerals are considered precious except to those who pry them from the mine, sift them from the tailings, swap them with other rock hounds, and display them proudly on their mantels.

Finally, we reach the core of the formation. Six-foot masses of quartz angle out of the ceiling. We have put on extra sweatshirts. It is 32 degrees Fahrenheit. We peer into the gloom and discover the reason. A massive slab of glacial ice lurks in the deepest recesses of the mine. It is but a temporary relief from the heat that slams into us as we clamber back to the upper world.

Later that evening, I visit the Frost Place. Scores of people huddle into Robert Frost's old timbered barn to hear poets who have traveled to this simple shrine to teach and read their poetry. The warm light shines out of the barn and people chuckle, grin, and give guttural little assents of approval when someone hits just the right word. An Irish poet, slightly inebriated on lite beer, finishes his reading just as an enormous tree crashes in the forest. Frost would have approved. I think, in fact, he pushed it.

A bear has pulled a hose out of the water pump at the poets' residence. The drought has made the bears feisty. The bear has made the poets thirsty. They sit around feisty, thirsty, and a little bit dirsty, discussing previous encounters with bears. The same conversation could have taken place in the Palermo mine or at Hubbard Brook. I have the same sense of being at a shrine, of being apart from the everyday world, of exchanging ideas with people who share the same odd passions and ideals. It doesn't really matter whether the practitioners are rock hounds, ecologists, writers, or poets. We share the same wellspring of awe inspired by nature. I ride my bike back home in darkness. Overhead, Perseid meteors remind me that summer is dwindling fast.

Summer is dwindling fast.

The Rockpile.

Chapter 20

The Rockpile

There may be worse weather, from time to time, at some forbidding place on the Planet Earth, but it has yet to be reliably recorded.

—William Lowell Putnam,
Defending Mount Washington's Weather, 1991

SEPTEMBER 3, 1999

It's early morning on the rockpile. I glance at my watch. 5:00 a.m. No sounds seep through the solid concrete walls of the Mount Washington Weather Observatory. No light filters through the thick-paned windows. The smell of garlic permeates my subterranean bunker. Last night we dined on whole cloves of elephant garlic drenched in Worcestershire sauce. Exotic dining is a tradition on the summit.

I'm lying on a simple wooden bunk made famous by the actor Ben Affleck, who slept here while filming a science program when he was only ten years old. The Observatory is built into solid bedrock. Its concrete walls are supposed to protect it from the hurricane-force winds that lash the summit on an average of twice per month. In the winter, this dorm is said

The giant orb of the sun rises majestically through the mountains.

to hum in the high winds, but last night it was enveloped in an ethereal, breathless silence.

I have journeyed to the top of Mount Washington to celebrate the end of the long, strange summer of 1999. But right now I'm feeling as isolated from the weather as if I were sleeping in a deep-sea submarine.

I can't sleep, so I might as well join the morning crew taking the first measurements of the day. I struggle out of my down sleeping bag rated for below-freezing temperatures. I put on my lined bluejeans, t-shirt, work-shirt, sweatshirt, and anarak, trying to remember the principles of layering. After all, it is 5:00 a.m. on Mount Washington, home of the worst weather on the earth. What do I find outside? The best weather on earth.

There is not a breath of wind. The temperature is 59 degrees Fahrenheit and rising fast. It is still dark enough to see the moon and a single planet. Last night, one of the observers saw a green fireball streak halfway across the sky. The crew can't wait for winter, when the mile-high curtains of the Northern Lights will shimmer across the evening sky.

I can still see clusters of tiny lights twinkling below me. They are villages nestled in the valleys between the craggy mountains of New Hampshire, Vermont, and New York. Reflections of the Atlantic Ocean loom beyond the lights of Portland, Maine, and the glow of Montreal and Boston can still be seen in the rapidly lightening sky.

There are no sounds save for Nin, the observatory cat, who yawns at me from Raven Rock. Nin's dominion over the summit is being weakly challenged by a wild fox, who is keeping his distance. Nin has been known to send rottweilers yelping back down the mountain after climbing up it all day.

Finally, the giant orange orb of the sun rises majestically through the mountains. Sparrows search for scraps of food and fat flies buzz in the breathless air. It feels like the beginning of a hot day at the beach. But the broad black back of Mount Madison looms in the middle distance and the desolate expanse of lichen-covered boulders reminds me where I am.

Soon the whistle of the cog railway will pierce the morning quiet. Hundreds of tourists will scamper over the mountain, snapping pictures and calling home. Few will notice the plaque that commemorates the 124 people who have perished on Mount Washington, the most deadly mountain in the lower forty-eight states.

Not far from the plaque is an endlessly running videotape of observatory personnel eating breakfast in sixty-mile-per-hour winds. Milk explodes out of its container, the observer is blown over backwards, and the table kites away in the wind. But a sixty-mile-per-hour wind is merely a fresh breeze on Mount Washington.

The observers have an exclusive club for people who can stay upright while walking across the observatory deck in 100-mile-per-hour winds. One of the observers who weathered Mount Washington's famous 231-mile-per-hour gust was pinned to the wall by the record-breaking blasts, unable to climb either up or down.

But I'm having a tough time imagining the worst weather on earth. I can

Mount Madison looms in the distance.

The first cog railway car up the mountain.

already feel the warm rays of the sun heating up the sultry air. In fact, today will be Mount Washington's hottest day of the year. The temperature will climb to sixty-nine degrees, setting a new record for September and almost breaking the all-time record high of seventy-two degrees.

The temperatures should be in the thirties at this time of year. I talk to a through-hiker who has just reached the top. He has trekked the entire length of the Appalachian Trail from Georgia to Maine without incident. But a few years ago, he almost froze to death on Mount Washington when a September blizzard shredded his tent and buried him under a foot of new snow. So what's going on here?

I peel off my carefully layered clothes—first the anarak, then sweatshirt, workshirt, and bluejeans. I'm down to my t-shirt and shorts, but I still feel hot. The only other time I was on Mount Washington was on June 12, when the summit set a heat record of sixty-seven degrees. In fact, Mount Washington has set eleven heat records this year and tied the record for the most snow falling in twenty-four hours: 26.9 inches.

And, right now, Mount Washington is in the midst of a six-day heat wave; the longest in its history. I'm trying to not take this weather too personally. But the only two times I have been to the top of Mount Washington the summit has broken heat records. Last week, when I drove to Montreal, the city was in a heat wave, with 90-degree weather that didn't cool all night.

Of course, someone arguing against global warming could say that none of these records are statistically significant. I would have to agree with them. But let's look at the summer of 1999.

The scene was actually set in 1998, when El Niño switched to La Niña in August. The Pacific Ocean had cooled, subtly shifting the jet streams north. Meteorologists predicted that this would cause more snow in winter, drought in summer, and more powerful hurricanes in autumn.

The drought came on gradually, as most droughts do. The last half of 1998 was drier than usual, but no one took much notice.

By March, we started to notice the first indications that something was afoot. New Hampshire newspapers sprouted headlines like, "Ice Fishing Meltdown" and "Whitewater Washout." Warm weather was melting lake ice earlier than usual and rivers were low from the lack of precipitation. The dry conditions had increased the pollen count by April and the threat of forest fires by May.

In June, New England had its first heat wave and the governors of New Hampshire, Massachusetts, Rhode Island, and Connecticut sent nonessential state workers home early to save on electricity. It was the first time that New England ever had to issue such a summertime electricity warning. The second was issued on June 28.

On July 6, New Hampshire set a new record. It switched from using more electricity to heat on a single day in winter to using more electricity to cool on a single day in summer. Remember, this is New Hampshire, not Florida. The record was treated as a curiosity, though, not a long-term trend.

If stories of the drought had taken on an Orwellian tone by July, by August they had switched to something you might have seen on reruns of "The Twilight Zone." Chicago had to hire a special refrigerated trailer to hold the extra bodies of people killed by the sweltering heat. Most of the corpses were of elderly people too poor to afford fans or air conditioning. In Dallas, an eight-month-old infant baked to death in a car while her mother spent the hot night drinking to cool off.

In New York City, thirty people died when a transformer blew because too many Manhattanites were using their air conditioners. Mayor Giuliani ordered suburbanites to rat on neighbors who watered their lawns. Ever resourceful, many New Yorkers simply bought artificial coloring to spray

In the midst of a heatwave.

Turkeys fared well in the drought, as they always do.

on their withering grass. In all, 215 deaths were attributed to the heat and sixteen states were declared disaster areas.

Even birds, bees, and bears were affected by the drought. Most birds laid fewer eggs because fewer insects were available to feed their chicks. They also congregated around water sources, incubating Saint Louis encephalitis, which killed three people and infected hundreds more. But some birds did well. Our neighbor counted twenty-three turkeys in a single flock. They had flourished in the dry areas abandoned by less-tolerant species.

Bees had worked overtime in the hot, dry spring. Now we are seeing the results of their hyperpollination. Old abandoned orchards are bent low with bumper crops of apples.

The apples are attracting bears who have had to roam farther afield to find food during the long, hot summer. Last week, I watched a mother bear herd three cubs across I-93. She raced them across one side, regrouped them in the median strip, raced them across the other side, then glowered at the traffic, daring anyone to follow. She is one stressed mother. I will give her a wide berth.

Of course, farmers have been hit the hardest. They have watched their crops wither, their topsoil turn to powder, their hard work and profits go up in dust. Most have been caught in a double whammy. It has cost farmers more to produce their crops, but prices will remain low because competing areas have not been affected.

Dairy farmers are a good example. A dairy farmer will typically own several hundred acres of land that he uses to grow hay and corn to feed his cows. Most summers, he will be able to make three hay cuttings. This year, he was lucky if he could cut his hay twice. He will have to buy more hay and corn to feed his cows and more fans and electricity to cool them off. They will still produce less milk, but the price of milk will remain low because it is regulated by the government. If New Hampshire is declared a disaster area, farmers will be eligible for low-interest loans. But most would prefer to "farm it out" rather than go further into debt.

But some farmers will use their disaster relief funds to install irrigation systems. In fact, most farmers prefer dry growing seasons to wet growing seasons, because they can control a drought but they can't stop the rain. But where will the water come from?

Many Canadians think that the United States will soon be eying Canada for its water. This summer, they passed a resolution banning the bulk export of water. The reason they gave for the popular ban is that if one sale of water goes through, water can be declared a commodity under the NAFTA agreement, and the United States could force Canada to sell more cheap water to the United States.

So the summer of 1999 has visited droughts, death, strife, and destruction upon us. It has turned old neighbors against each other. But will it convince us to do anything about global warming? I think not. The reason? A clue is chugging up the side of Mount Washington.

A big, black cloud of smoke signals that the cog railway is approaching. All I can see is the maniacal grin on the soot-covered face of the engineer.

One stressed mother. I will give her wide berth. Courtesy *The Courier* newspaper.

He pulls a cord and the shrill call of the steam whistle echoes over the mountain.

The fireman has had to shovel a ton of coal into the engine to push thirty-eight passengers up Mount Washington. Each passenger will be responsible for releasing 125 pounds of global-warming carbon dioxide into the air. The drivers who have driven up the toll road will be responsible for releasing 10 pounds of carbon dioxide per person. Even the admirable hikers who have climbed up the mountain on foot each have released 2.8 pounds of carbon dioxide into the atmosphere.

The implications are clear. Even if the entire population of the world decided to stay in bed and sleep, they would still produce too much carbon dioxide for our planet to absorb.

If I want to discover the reason for global warming, I have to look only as far as the nearest mirror. Six billion people on the earth want to eat, sleep, and do cool things like climb to the top of Mount Washington. A friend of mine has estimated that anytime we do anything as simple as buying a can of soup we are indirectly responsible for releasing ten pounds of carbon dioxide into the air.

In fact, the two most damaging things you can do to the environment are to make love and start your car. Unfortunately, these are also two of the most pleasurable things to do. It is part of our primate heritage. We have evolved dozens of neurotransmitters that fairly buzz with anticipation at the mere thought of a road trip or evening tryst. We have to fight two million years of successful evolution. That is the crux of the environ-

The last cog down the mountain.

The last car down the mountain.

mental crisis. That is the problem that most environmentalists fear to grapple with.

But now it is time to head back down the mountainside—time to face my own inconsistencies. How shall I get down? You guessed it. I will descend the mountain aboard that picturesque old anachronism, the Mount Washington Cog Railway. As Pogo so wisely put it, "We have met the enemy and he is us."

PART V

Autumn

Before the leaves can mount again
To fill the trees with another shade,
They must go down past things coming up,
They must go down into the dark decayed.

—Robert Frost, "In Hardwood Groves"

Autumn.

A male sports his first rack of antlers. Mass. Wildlife/Bill Byrne.

Chapter 21

The Baiting Season

OCTOBER 9, 1999

It is early dawn. Splashes of crimson and saffron slice through the dark green of conifers. High-flying geese carve perfect Vs across cobalt skies, and the morning mist carries the woody smell of rotting leaves. A lone raven sweeps low over the pine trees, quorks, and tumbles in mid-air. Is he trying to lead me somewhere?

Two moose emerge from the woods. The male sports his first rack of

The rut pit in our neighbor's beloved lawn.

bony antlers. He prances back and forth across our neighbor's lawn. He seems to be irritated by their newly planted maple. He sweeps his antlers through the leaves, breaking branches and shattering the trunk of the fifteen-foot tree. The female saunters off as if disinterested, but positions herself so that she can continue to watch his show-off behavior.

The male trots to the center of the lawn and paws heavily with his front hooves. Soon he has excavated a nine-foot rut into the beloved lawn. The female vocalizes quietly, but does not move. The male stretches his body over the pit and urinates. The female gallops to his side and circles him while mooing loudly. The male straightens up. The female shoves him out of the pit and drops on all fours to rub her neck and shoulders in the rut. She rolls and wallows in the heady mix of urine, mud, and pheromones.

The male is now lying near the pond, watching her intently. Duly perfumed, the female approaches the male and fondles his new antler with her muzzle. The young male's ears flap with excitement, and the two trot mincingly back into the woods. Tomorrow, our neighbor will gain a new appreciation of the rowdy nature of teenage moose.

The female approaches the male. Courtesy *The Courier* newspaper.

Not far away, a mother bear moves purposefully through the underbrush. Her cubs follow haphazardly behind. One finds a nest of yellow jackets flying in and out of a small pine. The tree was bent to the ground and shattered by a winter storm. The cub stops to slap at the insects. He traps two under his paw and licks them off his palm. He is the curious cub, always loitering behind to turn over rocks or dig up anthills.

The cubs have gained several inches of fat since they changed their diet from the sugary berries of summer to the fattier foods of autumn. You can see their paths emerging from the deep woods to the edge of abandoned orchards and cornfields. You can see where they have dragged the cornstalks

A young moose flaps his ears with excitement. Mass. Wildlife/Bill Byrne.

Tree damaged by rutting moose. Tomorrow our neighbor will gain a new appreciation of the rowdy nature of teenage moose.

back into the forest. Their scat is now filled apple seeds and pin-cherry pits. The trees are left with dead and broken branches.

Now the mother bear has reached a ridge with beech trees. The sun shines on the trees' smooth gray bark and the cubs scuff through the crunchy dry leaves. The mother bear clambers up a tall beech, leaving a trail of fresh claw prints in the smooth bark. The tree is covered with the scars of other bears who have climbed this tree for forty years. She balances in a crotch of the tree and stretches far out to grab great armloads of branches. She pulls them through her teeth to dislodge the nutritious young beechnuts.

Many of the nuts clatter to the ground where the cubs are snuffling through the leaves like boars after truffles. The curious cub lies on his stomach in a bed of the sun-warmed leaves. He is trying to balance a single beechnut on his wrist so he can crack it open with his teeth. The others simply wolf down the nuts.

After the mother shreds the branches, she places them under her ample rump. By the end of her feeding, she has accumulated a mass of bent and broken branches that look like a giant eagle's nest. Long after the other leaves have fallen, this tangle of matted vegetation will remain in the tree to confound curious birders.

The mother snoozes quickly in her "bear nest" while her cubs finish up the scraps below. Her behavior has allowed the cubs to gorge on this important food source long before it has been consumed by deer and turkey. They must await for the nuts to fall of their own accord.

Normally, the beechnuts would provide the bears with a large percentage of the fat so crucial to their successful hibernation. But this year the nuts are smaller and fewer than usual. Perhaps it was because of the long dry summer.

But the female knows of another source of food. She climbs down the

The curious cub. Courtesy *The Courier* newspaper.

Bear feeding in tree. Mass. Wildlife/Bill Byrne.

Male black bear. Charles H. Willey ©.

tree and grunts at her cubs to follow. The smallest struggles behind. Further down the mountain, the mother smells another bear. It is a young male. She grunts at her cubs, who scamper up a nearby tree and peer through the leaves, slapping the branches nervously.

The female rises on her back legs and huffs at the intruder. He might be one of her own offspring from years before. He is lean and hard working, one of the well-mannered bears of Coal Hill, not one of the nuisance bears that hang around the dump.

The female drops on all fours and charges, slapping the vegetation as she comes. She stops in front of the startled juvenile and slams a nearby tree. The young bear runs off, bellowing.

Now the female pauses to smell the air. It is what she has been expecting. She stops at a clearing, looks around, and carefully leads her cubs between two fallen trees. The carcass of a dead calf sticks out from under one of the logs. It will fill her cubs for several days and provide fat for their hibernation. She grasps the hair on the calf's stomach and tugs. Warm entrails spill onto the newly fallen leaves.

The hunters.

One of the cubs charges to her side. As the mother turns to grunt at his mischievous behavior, a searing pain radiates through her neck. She rears to grab at the arrow, but another enters her belly. She yelps in pain and anger. The cub has also been hit. She staggers to his side and bites at the arrow sticking from his flank. The cub snaps at his mother, wondering why she has turned on him.

Another arrow flies and the mother falls. Her body goes limp and blood gushes from her open mouth. Not far away, the other two cubs bleat with fright and run back hesitantly into the woods.

Whoops of joy ring from the trees overhead. Two hunters climb down from chairs they have bolted to the pine trees overlooking the site. The hunters have spent the last three weeks carefully training the bears to come

The dead.

The blind.

to the bait. They have hauled carcasses from nearby slaughterhouses. They have heated honey over a propane stove so that its sweet white smoke would drift through the forest. The smoke has stuck to the trees, leaving a sweet trail back to the site.

The hunters have buried frying oil in front of the carcasses so the bears will track its odor back into the woods. The hunters have hung their worn t-shirts near the bait to accustom the bears to their human scent.

It has been hard work, but it has all paid off. Both men have killed their bears on the first day of the season. It has been about as easy as running over pedestrians in a walking zone—and legal too, in Maine and New Hampshire.

The October sun filters through shimmering layers of birch, beech, and sassafras leaves at the Hubbard Brook Forest.

Chapter 22

The Experiment

OCTOBER 19, 1999

It is late afternoon at the Hubbard Brook Experimental Forest. The October sun filters through shimmering layers of birch, beech, and sassafras leaves. The woods are suffused in a warm, buttery glow. It is as if the forest is experimenting with every luminescent combination of yellow—saffron, amber, celandine, gold.

This is the peak of translocation, that bittersweet time of year when cold nights kill chlorophyll, stop photosynthesis, and rip masks off the accessory pigments that have lain hidden within the leaves all summer. This year, the movement of nutrients into stems reveals something else. The yellow leaves of beech and birch are luminous and vibrant, but the leaves of sugar maples are dull and dry.

I pluck some leaves to compare the two. The beech leaf is stiff and waxy, the maple dry and curled. Are beech trees better at drawing water through their roots? Are they better at slowing water from escaping through the pores in their leaves? Have beech trees retained water better during the long, dry summer? Has the mild autumn not killed off the chlorophyll as quickly as in years past?

Birch trees exploit a new patch of sunlight.

A fallen ash provides some clues. The fifty-foot tree toppled in this summer's July 6 microburst. Now a gaping hole in the canopy allows sunlight to flood through to the forest floor.

Yellow birch will be the first trees to exploit this new sunny niche. Theirs is a profligate strategy. They produce millions of small light seeds that blow through the forest, enabling them to quickly exploit new patches of sunlight. As the birches grow, however, they will create a new sundappled niche, unfit for their own seedlings, but ideal for sugar maples.

Sugar maples have a moderate strategy. They produce a moderate amount of "helicopter" seeds that travel moderately well and have a moderate amount of nutrition to sustain them when they land.

The last to arrive, decades later, will be the shade-tolerant beech trees. Theirs is the most conservative strategy of all. They produce a small number of large, heavy nuts that normally drop to the ground in the deep shade of their parents.

But beech trees also reproduce by sprouting from roots. So this copse is actually a colony of beech tree clones that share common roots. Their sprawling network of roots is able to supply the trees with water during the driest of years. So, during times of stress, the pioneering strategy of birch and the stable strategy of beech would seem to have the advantage over the moderate strategy of sugar maples.

But what is causing the stress? I stoop to inspect a maple seedling. It has sprouted as usual, but now appears unhealthy. In many places, sugar maples

Chris Johnson shows the distinct white wash that delineates the organic soil above and the mineral soil below. Acid rain has depleted soils that have taken thousands of years to form.

are being displaced by red maple, trees less valuable for lumber and less arresting as foliage. So, what else is going on in this forest?

Chris Johnson may have part of the answer. He is a biogeochemist from Syracuse University who has been studying soils at Hubbard Brook for over a decade. Right now he is kneeling in a shallow pit, cleaning debris off a soil profile—a profile so perfect it deserves to be in a textbook. A distinct wash of white clearly delineates the organic soil above from the mineral soil below. This is the chemistry lab where anions and cations are exchanged.

The organic soil is where leaves decompose, releasing natural acids, and where rain makes base cations of calcium available to trees for growth. But

This forest stopped growing in 1987 and has not gained biomass since.

The Huey helicopter arrives.

sulfates in acid rain are inorganic. They do not decompose like natural acids, but seep down into the mineral horizon, where they leach calcium from the soil.

Biogeochemists like Chris Johnson suspected that this loss of calcium might nullify the effects of removing sulfates from acid rain. But they weren't sure until they compared notes with their bucket biologist colleagues. This time, the bucket biologist was Tom Siccama from Yale's School of Forestry and Environmental Studies.

Tom has spent the past thirty years measuring the growth of every tree in this thirty-acre watershed of forest. He has weighed fallen leaves and twigs. He has measured the diameter of every tree at breast height. He has drilled cores into select trees so he could count their growth rings and measure the amount of calcium in their inner and outer wood.

Filling the hopper.

The result? He made an astounding discovery. This forest stopped growing in 1987 and has not gained biomass since. In fifty short years, acid rain has depleted soils that have taken thousands of years to form.

Gene Likens, head of the Hubbard Brook Research Foundation, explains that the problem is really twofold. Not only has acid rain leached calcium out of the soil, but the Clean Air Act has removed calcium from the atmosphere.

Forests like Hubbard Brook used to receive half of their calcium from the weathering of bedrock and half from depositions of atmospheric dust. Prior to the Clean Air Act, atmospheric dust was produced in large quantities by such things as wind erosion, mining, and the manufacturing of cement. Even automobiles driving over unpaved roads were a source of calcium. The Dust Bowl era must have been a particularly good time for aerial dispersal of calcium—now there is the germ of an idea.

Obviously, the way to solve the problem was not to gut the Clean Air Act; atmospheric dust also causes severe health problems. No, "Hubbard Brookers" proposed an unprecedented, some would say audacious experiment—unprecedented in that they would use a helicopter to spray fifty tons of calcium over thirty acres of woods; audacious in that it would take fifty years to see the results.

That experiment is taking place today. A shiny gray Huey helicopter is making a series of sorties overhead. I feel like a Viet Cong photographer stooping low to protect my camera from the carpet of pelletized calcium that it is raining down on the canopy above. The tiny, BB-sized pellets are bouncing off the leaves, landing on the litter, rolling into the organic soil.

Setting out mylar balloons to guide the helicopter.

A Huey helicopter spreads 55 tons of wallasonite onto the watershed, to replace the calcium leached away by acid rain.

The calcium is in the form of wollastonite, a calcium silicate similar to that found naturally in the bedrock around Hubbard Brook. But this wollastonite was mined from a talc mine in upstate New York, pelletized at a fertilizer plant in Illinois, and delivered two days ago by three tractor-trailer trucks. It represents as much calcium as has been removed from this watershed in fifty years.

But all is not well. The lawnmower-sized engine that disperses the pellets has been jamming. Ian Helm has led a crew of foresters to a clearing near the top of the ridge overlooking the watershed. Each time the pilot returns with another ton of calcium, he hovers over the ridge so that Ian can take a few well-placed whacks with a sledge hammer.

Eventually the recalcitrant motor starts. It seems somehow reassuring

Researcher weighs and measures leaves treated with calcium.

Helicopter departs.

that such a low-tech solution still has a place in an experiment guided by satellite technology and the blue mylar party balloons we threaded carefully through the trees to delineate the site.

By the end of the day, the sun has set and the helicopter has had to switch on its lights to return to Concord. All but nine tons of the calcium have been spread on the watershed. They will be dispersed when the weather breaks in a few days.

Don Busso has coordinated students into six-hour shifts so they can monitor the weir at the bottom of the watershed every hour for twenty-four hours. The pH in the stream has already risen from 5.2 to 5.8, about the acidity of rain water. A slight film on the surface of the water indicates that the binder that holds the pellets together is breaking down. The water has taken on the color of very light tea with just "*un petit nuage du lait.*"

The leaves in the forest are covered with tiny white patches. The carbon is breaking down, leaving the slow-release calcium behind. Soon the snow will fall and the calcium will be sealed below the snowpack.

When the snow melts in the spring, the calcium will percolate into the soil and the experiment will begin. It will take seven years for all the calcium to seep into the soil. Scientists and students will be on hand to collect roots, algae, bacteria, and fungi; mites, snails, spiders, and insects. They will use isotopes of strontium to trace the calcium as it flows through the system.

If all goes well, they will see a cascade of effects in soils, streams, and trees. They will see an increase in species richness and overall productivity. And, if in fifty years, sugar maples dominate the succession of this forest,

"Brookers" will know that their hypothesis was correct, forests can be medicated back to health after the effects of acid rain.

In the meantime, however, they will have discovered something far more fundamental. They will have learned more about the profound yet subtle ways that calcium, just one of the macro-nutrients that flow through our biosphere, make life and forests possible.

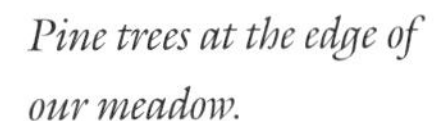

Pine trees at the edge of our meadow.

Chapter 23

To Fell a Tree

> *Fragrant little chips of history spewed from the saw cut, and accumulated on the snow before each kneeling sawyer.*
>
> —Aldo Leopold,
> *A Sand County Almanac*, 1949

NOVEMBER 13, 1999

It started as a simple problem. A pine tree on the edge of our meadow had grown too large. Its crown was casting deep shadows across the grass and its branches were obstructing our view. Our pasture had become a long, dark alleyway only bathed in sunlight a few hours out of the day.

First, I tried to balance a ladder against the tree and lop off its protruding branches. But I could reach only so many branches. Then I decided to saw down the tree to open a swath of light across the darkened meadow.

It would be a moderate undertaking. The old pine was twenty-four inches in diameter. I calculated where I wanted the tree to fall, and revved up my chainsaw. This tree was too big for my handsaw. Besides, this was not designed to be a long poetic experience, but a quick prosaic chore.

I measured my notch so the tree would fall neatly on the meadow, and made my first incision. The saw bit quickly through rings of lignin and the notch fell gently to the ground. I moved to the other side of the tree and started my back cut. But the saw would penetrate only so far. The manufacturer had put a guard on the tip of the bar so it could only tackle a sixteen-inch tree.

I rocked the saw back and forth, to no avail. The blade would not penetrate deep enough so that the tree's weight could crack its spine. I started another cut, higher than the first, but it did no better. I tried to deepen the notch but feared the saw would get bound in the sawkerf of the toppling tree. I tried the axe, but it was long, slow labor.

Gradually it dawned on me. I had bitten off far more than I could chew. I couldn't leave the tree hanging, ready to blow over in the slightest wind. I wished for a propitious breeze, but none came to save me from my hubris.

I placed a sheepish phone call to a neighboring forester to see if he had a larger saw. During the embarrassing call, I heard a series of splintering cracks and watched the great tree fall with a majestic, earth-shattering thud.

But the die had been cast. The forester came, and we surveyed the rest of the trees. If we removed the first line of trees, the meadow would be thirty feet wider and we would still have the second line of pines along its edge. I asked how much it would cost. He calculated his time, machinery costs, board feet, and came up with his answer.

"Oh, I expect you'll make a few hundred dollars on the deal."

"What?"

I had become a timber baron overnight. But money was just the tip of the iceberg. The trees would provide a few days' work for my neighbor. They would provide jobs in the neighboring town's sawmill.

I will miss that old tree.

My trees heading south to build somebody's new home.

Removing the trees would provide more light for the pasture to grow, more grass for the deer to graze on, more raspberries for songbirds, turkey, and bears.

But I will miss those old trees, and I thank them for their sacrifice. For fifty years, they have been sweeping carbon from the atmosphere and sequestering it in their inner wood. They have been reducing a greenhouse gas, helping to keep our planet cool, helping to keep our earth habitable for life.

Their lumber will go into some young couple's new home, keeping its carbon out of circulation. Their slash will decompose back into the soil to regenerate new trees. As long as I don't burn the trees' wood, their carbon will stay sequestered.

The trees represent part of the five acres of forest I need to offset the amount of carbon dioxide I produce as an average American to heat my home, drive my car, and power my luxuries and necessities.

In 1850, the average New England farmer needed ten acres of forest to supply the firewood to heat his home and another few acres of hayfields to feed the animals that supplied his transportation. I suppose that represents some kind of index of efficiency even as we gobble up the world's remaining non-renewable fossil fuels.

In fact, these resilient New England forests represent a valuable world asset. New Hampshire has enough forest cover to offset all the carbon dioxide produced by her inhabitants, with enough left over to support two hundred thousand out-of-staters, or as my neighbor summed it up, "So you mean New Hampshire supports two hundred thousand Massholes?"

Some economists have put it in more statistical terms. They have calculated that every twenty acres of forest we cut in New England represents an acre of forest "saved" in the tropics. They argue that this is good for the world because New England's forests can be harvested renewably, but once a tropical forest is gone it is gone forever. This is because tropical forests

Reflections. I hope I have made a good and wise decision.

hold most of their nutrients in their leaves rather than in their soils. After a tropical forest is cut, its nutrients are removed and its soil turns into laterite, a dry, hard soil with the consistency of baked clay. A new forest cannot get started.

So these pines have given me another gift. They have brought me into their world, made me think globally while acting locally. I hope I have made some good and wise decisions.

Chapter 24

Senseless in Seattle?

Think Globally, Act Locally —An old environmental saying

NOVEMBER 1999

"Look at those crazy bastards!"

The television is on at the Village Inn bar. We are watching people dressed as sea turtles cavorting in front of the cameras. They are protesting against the World Trade Organization in Seattle.

Normally, at this time of year, the television would be tuned to a ski event. Two years ago, all of Franconia watched as a favorite son competed in the Winter Olympics.

But this year, nobody wants to think about skiing. It's just too damn depressing. Last summer, Cannon spent six million dollars to build new lifts and trails, but the mountain is still closed. It is still warm and snowless; too mild, even, to make artificial snow. No one can remember the last time the ski season got off to such a lousy start.

I've come to the bar to think globally. Not a bad place, as it turns out.

"Who would've thought a bunch of lefties would protest the World Trade Organization?"

"Yeah, you know, I can still remember my economics professor trying to pound it into our little heads that anything that helps free trade is good and anything that hurts free trade is bad. He used to speak at a lot of rotary clubs, and always implied that if you were for free trade you were good and liberal and if you were against it you were simply small town, conservative, and bad. Now these guys are turning everything upside down."

"Well, at least they've managed to kill that old saw, 'Jobs Versus the Environment.'"

"Oh, you two are just a bunch of pinko Luddites!"

"Luddites, give me a break. That was the eighteenth century. It's a different world now. We've had two hundred years to fill the air with greenhouse gases; two hundred years to wash away topsoil; two hundred years to clearcut forests; two hundred years to use up oil, gas, and metals. The economic world is already consuming resources faster than the natural world can produce them, and every time a forest or fishing ground is wiped out we lose more jobs."

"But what about new technology? Look at e-commerce and the Internet."

"Oh sure, the world is caught up in a speculative frenzy over the virtual world, but that's just become a way of selling more things faster. Anyone who works with natural resources knows what you don't want is quick profits, what you do want are long-term jobs."

"So what do dancing turtles have to do with jobs?"

"Okay, the networks are doing a great job of making it all seem ridiculous. But what's really at stake are things like forests. Some countries have tariffs as high as 40 percent to protect their local forestry industries. If the WTO trades those away, you'll see incredible pressure to chop down forests in the Pacific Northwest, New England, Indonesia, and Brazil.

"Sure, you'll see a few big, fat international companies make huge profits, but then there'll be nothing. No work left for people who have lived off these forests for hundreds of years."

Overnight, a million acres of forest that had produced paper, lumber, and jobs for five generations of New Englanders were converted into second homes.

Christmas trees or firewood? Courtesy *The Courier* newspaper.

"But won't globalization mean New England will make more money selling lumber to China and Japan?"

"Are you kidding? We've already had a glimpse of globalization. You know the Diamond Company?"

"You mean the company that makes those cute little match boxes?"

"Yeah, that's the one. But they also make paper and lumber. For years, they owned over a million acres of forests in Maine, New Hampshire, and Vermont. They cut them wisely to keep their mills busy. But with all those assets they were a sitting duck."

"Ready to pluck."

"Yeah, ready to pluck. Anyway, along comes this British guy, see—a corporate raider type. He buys out the company, a hostile takeover. He could care less about forests, jobs, or lumbering."

"So what does he do?"

"What does he do? What would you do? He's spent all his money and just wants to make it back fast. First he tries to clearcut the land for a quick profit, then he realizes he can unload it to another European Company. What was its name? Some kinda French name."

"CGE, Cie General Electricité, a communications company."

"Whatever, they turn around and sell the land to a Boston company that puts it up for sale for vacation homes. 'Boom,' overnight a million acres of

Firewood. Courtesy *The Courier* newspaper.

forests that produced paper, lumber, and jobs for over five generations are converted to condos. That's a million acres and jobs in Maine, New Hampshire, and Vermont. That's globalization for you. It's not about what's good for the globe, it's about what's good for some fat-assed businessman who wants to cut and run."

"I don't think globalization is really the problem. The problem is that we are letting companies define globalization in their own greedy, self-interested way. We should be making global decisions but they should be based on ecology as much as on economics."

"What the hell do you mean?"

"An ecological globalist should look at the world the way a forester looks at a woodlot. First he tests the soil and measures the rainfall to determine whether the land is best suited for hardwoods or softwoods. Then he factors in economics to decide whether the forest should be used for logging, maple sugaring, growing Christmas trees, or simply for firewood. Finally he comes up with a long-term plan to ensure that the land keeps producing the kind of trees that it is naturally inclined to support.

"If economists started to think that way, they might decide that wheat fields should remain wheat fields and not be converted to suburbs. They might decide that the North Woods are more valuable as reserves for lumber while tropical forests are more valuable as producers of oxygen. They might decide that land that supplies basic necessities such as food, water, oxygen, and trees should be kept for the future and not exploited for the quickest buck. The way to do that is often through things like tariffs and taxes."

"Hey, enough already with global thinking. It's starting to make my head hurt. What about acting locally? What about this idea of letting 180 square miles of the White Mountain National Forest grow into old-growth forest? Doesn't that run counter to your global thinking?"

"Not at all. Look at your forest up there on Coal Hill. People have owned and worked that land for hundreds of years. They burned it for charcoal, logged it for lumber, cut it for firewood, and farmed it for food.

"Despite all this, the forest has returned as resilient as ever. Now the hill is a variegated pattern of old-growth forest, white pine stands, abandoned fields, and rejuvenating logging areas. That mix of different ecosystems supports a greater variety and abundance of plants and animals than any could alone. It's the same with forests throughout northern New England.

"So now we have this incredible chance to let a pretty significant piece of woods grow into a mature old-growth forest. That's the scarcest ecosystem in the mix. Most of New England's forests are less than seventy years old, just post-pubescent in forest years.

Logs to be chipped for fuel to produce electricity.

"That old-growth forest can provide things that none of the other ecosystems can provide, like unfragmented habitat for large forest animals."

"Oh great, so we'll get mountain lions back in the woods!"

"Maybe."

"And of course it's easy to do, because it's already public land."

"So it shouldn't cost tax payers any money."

"Shouldn't, and it won't really affect the lumber industry because only 4 percent of the logging done in the state is done in the National Forest anyway.

"But old-growth forest is really just part of the mix. A true globalist would want to maintain that healthy mix of forest types by building on a natural coalition of loggers, farmers, woodlot owners, and environmentalists."

"Natural coalition?"

Old-growth forest.

Pileated woodpecker. The result of reforestation. Charles H. Willey ©.

"Hey, nobody said it would be easy. But if they work at it, they can ensure that the forests of northern New England continue to grow trees to supply oxygen for the world, livelihoods for forest workers, and places of peace, solitude, and rejuvenation for recreationists. That would be the true meaning of globalization in my book."

"I'll drink to that."

Mysteries of the still-unfrozen pond.

Chapter 25

My Father's Ashes

Before the leaves can mount again
to fill the trees with another shade,
They must go down past things coming up,
They must go down into the dark decayed.

—Robert Frost, "In Hardwood Groves"

NOVEMBER 28, 1999

It is late November, but the weather belies the season. There is no wind and the temperature hovers just below freezing. Flakes of snow drift quietly from scattered clouds, but the crusted flanks of Lafayette are bathed in occasional sunshine.

I watch the snowflakes land as I trudge along the now-familiar path. I watch them disappear into the black mysteries of the still-unfrozen pond.

It is November 28, the anniversary of the day I first entered these woods to write this book. Now I'm returning, a little smarter perhaps, perhaps a little wiser.

I circle round the pond and enter the quiet woods. Now I can hear the snow as it patters on the forest floor. The wind whispers through the conifers above and the sunlight filters through a nave of rustling needles.

An antler lies in the path, probably left here by a coyote. It is small and

puny; gnawed on a bit by mice, but nonetheless a sign. The moose have returned to their wintering grounds. They have started to feed on twigs and bark.

An antler lies in the path.

I pick up the antler and continue into the moose yard. Here the forest floor is an earthen tapestry of lichen, moss, needles, and fern. A few piles of scat lie on velvety pillows of deep green moss, pillows delicately embroidered with flakes of perfect snow.

Piles of scat lie in velvety pillows of moss and lichen.

I scatter my father's ashes around the antler.

Now the snow is falling faster. It forms a veil of shimmering white. A maple tree emerges. Its trunk is scored with fresh wounds from the teeth of browsing moose. This is a tree that can use a little help.

I prop the antler against the maple and remove a small plastic bottle from my jacket pocket. The wind rises in the trees overhead, the snow settles on the forest floor. I kneel in the soft moss and scatter my father's ashes in a perfect circle around the trunk of the ailing tree.

This is resurrection I can believe in, reincarnation I can understand. Calcium from my father's bones will revitalize this soil, soil leached by acid rain. It will replace this tree's calcium, calcium browsed by a hungry moose. Next summer my father will live again in the maple's new green foliage.

Indians think the lenticels of birch trees are the eyes of our ancestors.

Watching over me in the forest. Charles H. Willey ©.

Perhaps my father's spirit will reside in these woods. Perhaps the lenticels of birch trees are the eyes of ancestors watching over us in the forest. The Indians thought so. I don't know, but I know I will never think of them the same way again. I don't know, but I know my father's memory will live whenever I think of this spot.

Suddenly a branch snaps and a large creature gallops unseen through the woods. I swirl around. My heart is racing. Perhaps I will see the bull moose that has eluded me all summer. I click in my telephoto lens and set out in hot pursuit.

But after a few steps I stop. Perhaps it is better that I don't see the old moose. Perhaps it is better I just remember him watching over me in the forest.

Index

Preparing for winter.